Aerospace Science and Engineering
IV Aerospace PhD-Days

IV Aerospace PhD-Days
International Congress of PhD Students in
Aerospace Science and Engineering
Scopello, 6-9th of May 2024

Editors
Andrea Alaimo[1], Antonio Esposito[1] and Marco Petrolo[2]

[1]Università di Enna Kore, Italy
[2]Politecnico di Torino, Italy

Peer review statement

All papers published in this volume of "Materials Research Proceedings" have been peer reviewed. The process of peer review was initiated and overseen by the above proceedings editors. All reviews were conducted by expert referees in accordance to Materials Research Forum LLC high standards.

Published under License by **Materials Research Forum LLC**
Millersville, PA 17551, USA

Published as part of the proceedings series
Materials Research Proceedings
Volume 42 (2024)

ISSN 2474-3941 (Print)
ISSN 2474-395X (Online)

ISBN 978-1-64490-318-6 (Print)
ISBN 978-1-64490-319-3 (eBook)

This book contains information obtained from authentic and highly regarded sources. Reasonable efforts have been made to publish reliable data and information, but the author and publisher cannot assume responsibility for the validity of all materials or the consequences of their use. The authors and publishers have attempted to trace the copyright holders of all material reproduced in this publication and apologize to copyright holders if permission to publish in this form has not been obtained. If any copyright material has not been acknowledged please write and let us know so we may rectify in any future reprint.

Distributed worldwide by

Materials Research Forum LLC
105 Springdale Lane
Millersville, PA 17551
USA
https://www.mrforum.com

Manufactured in the United State of America
10 9 8 7 6 5 4 3 2 1

Table of Contents

Preface

The Aerospace PhD Days are organized by the Italian Association of Aeronautics and Astronautics, AIDAA, and are open to PhD students working on Aerospace Science and Engineering topics. The first two editions were held in Pisa in 2018 and 2021 under the supervision of Professor Aldo Frediani; the third edition was held in Bertinoro in 2023. The event has no parallel sessions to allow students to follow all presentations; furthermore, most of the organization is carried out by graduate students.

The 2024 edition had 45 presentations, with students from more than ten institutions, including delegates from China, Germany, Lithuania, and Switzerland. Many aerospace disciplines and topics were covered, such as fluid dynamics, structures, stratospheric balloons, maintenance and operations, UAV, dynamics and control, space systems, sustainability of aeronautics and space, aeroelasticity, multiphysics, space debris, aeroacoustics, navigation and traffic management, additive manufacturing, and human-machine interaction.

Three invited speakers delivered lectures on opportunities, challenges, and perspectives in the aerospace field: Angelo D'Agostino (Head of Research Career and NCPs coordination Uni, APRE); Marco Brancati (Chief Technology & Innovation Officer, Telespazio Group); and Riccardo Augello (Postdoctoral Fellow under the Marie Sktodowska-Curie Actions, Politecnico di Torino and Caltech).

Roberto Galatolo, Università di Pisa, Pisa, Italy
Tomasz Goetzendorf-Grabowski, Warsaw University of Technology, Warsaw, Poland
Gianluca Iaccarino, Stanford University, Stanford, CA, USA
Mohamed Ichchou, École Centrale Lyon, Lyon, France
Lukasz Kiszkowiak, Military University of Technology, Warsaw, Poland
Michèle Lavagna, Politecnico di Milano, Milano, Italy
Enrico Lorenzini, Università di Padova, Padova, Italy
Salvo Marcuccio, Università di Pisa, Pisa, Italy
Roberto Marsilio, Politecnico di Torino, Torino, Italy
Francesco Marulo, Università di Napoli Federico II, Napoli, Italy
Pierangelo Masarati, Politecnico di Milano, Milano, Italy
Marco Montemurro, École Nationale Supérieure d'Arts et Métiers, Bordeaux, France
Fabio Olivetti, Università di Bologna, Bologna, Italy
Fabrizio Oliviero, TU Delft, Delft, The Netherlands
Francesco Picano, Università di Padova, Padova, Italy
Sergio Ricci, Politecnico di Milano, Milano, Italy
Lorenzo Trainelli, Politecnico di Milano, Milano, Italy
Mauro Valorani, Università di Roma La Sapienza, Rome, Italy
Wenbin Yu, Purdue University, West Lafayette, IN, USA
Navid Zobeiry, University of Washington, Seattle, WA, USA

Organizing Committee

Andrea Alaimo, Università di Enna Kore
Sara Bagassi, Università di Bologna
Francesca Bracaglia, Politecnico di Torino
Marzia Corsi, Università di Bologna
Antonio Esposito, Università di Enna Kore
Nunzia Favaloro, CIRA
Vincenzo Gulizzi, Università di Palermo
Giuseppe Mantegna, Università di Enna Kore
Pierangelo Masarati, Politecnico di Milano
Rebecca Masia, Politecnico di Torino
Dario Modenini, Università di Bologna
Martino Carlo Moruzzi, Università di Bologna
Marco Petrolo, Politenico di Torino
Elisa Tortorelli, Politenico di Torino

Aerospace Science and Engineering - IV Aerospace PhD-Days
Materials Research Proceedings 42 (2024) 1-4

Materials Research Forum LLC
https://doi.org/10.21741/9781644903193-1

CFD analysis of a hydrogen powered scramjet with multistage injection

O. Russo[1,a*], M. Marini[1,b], P. Roncioni[1], F. Cascone[1], S. Di Benedetto[1],
M. Albano[2], G. Ranuzzi[2]

[1] Italian Aerospace Research Centre (CIRA), Via Maiorise snc, 81043 Capua, Italy

[2] Italian Space Agency (ASI), Via del Politecnico snc, 00133 Rome, Italy

[a]o.russo@cira.it, [b]m.marini@cira.it

Keywords: Scramjet, Hypersonic, CFD, Supersonic Combustion

Abstract. The PhD topic focuses the attention on the internal flow path analysis of a hypersonic demonstrator equipped with a scramjet engine, which will fly in a 10-second experimental operation. The work is part of a project funded both by the national program PRORA, and ASI, by means of an agreement between CIRA and ASI on hypersonics. The combustion is non-premixed and the fuel chosen is hydrogen which is injected by multistage strategy in two different stages. Engine performance and parameters such as combustion efficiency are analyzed through nose-to-tail CFD analysis both in fuel-off and fuel-on conditions. An important result is the achievement of a high combustion efficiency by the conclusion of the engine cycle with stochiometric hypotheses and single step reaction. Future studies will examine the performance of the scramjet using more detailed chemical models and varying the injection strategy while also evaluating the effect of the air-fuel ratio. Finally, the numerical results will be possibly validated with experimental measurements.

Introduction

In recent years, several research projects supported by the European Commission have been launched with the aim of developing a high-speed aircraft that simultaneously combines aerothermodynamic, structural and propulsive aspects. CIRA, taking advantage of the strong involvement and experience in European projects such as HEXAFLY-INT [1] (flight test of a vehicle without propulsion for hypersonic flight), and previously in HEXAFLY [2], has set the challenge of designing a scaled hypersonic scramjet demonstrator for a future test flight, named the *Scramjet Hypersonic Experimental Vehicle* (SHEV). The project is co-funded by ASI.

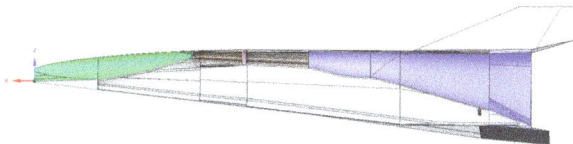

Fig. 1. Internal flow path of demonstrator. Inlet (green), combustion chamber (brown), nozzle (violet), combustor's struts (pink).

The scramjet engine is integrated into the demonstrator and is constituted by intake, combustion chamber and nozzle. The combustor is equipped with a two-stage multi-struts injector system, that is composed by two semi-struts, located at the beginning of combustor, and a full-strut located at the central zone of the same. The semi-struts distribute 65% of the hydrogen mass flow rate to the external regions; the central full-strut disperses the remaining quantity of fuel. As shown in Fig 2, the mission scenario involves an air-launched solution with an aircraft carrier releasing (Sep I) a launch system, consisting of a hypersonic demonstrator and its launch vehicle propelled by a solid booster, which brings the payload at specific speed and altitude. The booster accelerates the demonstrator to hypersonic speed through a controlled trajectory and releases it for a 10-second experimental operation (Sep II). CFD analysis employed the free stream conditions chosen for the experimental window (Mach 6÷8 and altitude 27÷32 km) to understand how the demonstrator behaves in these horizontal level flight conditions.

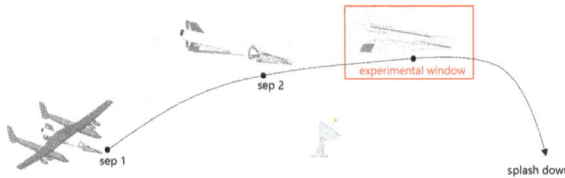

Fig. 2. Mission scenario

Numerical analysis and results

The performance evaluation of the scramjet engine was conducted by means of Computational Fluid Dynamics (CFD) analyses of the internal flow path. A simplified configuration was chosen focusing on half of the setup and employing an unstructured grid consisting of 1.9-million cells.

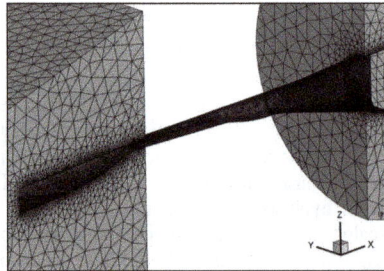

Fig. 3. Computational grid for simulation

Preliminary numerical analyses were performed with inviscid flow and in fuel-off conditions to assess the air MFR (mass flow rate) under various conditions chosen for the experimental window. The single-step chemical scheme (eq. 1) with a stoichiometric air-fuel ratio, $\phi=1$, was assumed to model air-hydrogen combustion. This approach made it possible to determine the total hydrogen MFR, that is partitioned and injected into the combustor with a multi-stage injection strategy.

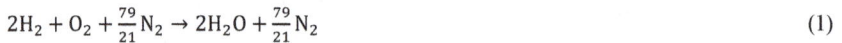

$$2H_2 + O_2 + \frac{79}{21}N_2 \rightarrow 2H_2O + \frac{79}{21}N_2 \tag{1}$$

Further analyses have been performed to evaluate the effects of viscosity on internal propulsive flow path performance of SHEV with k–ω SST model. The quantities shown in the following figure represent the average values, integrated over different sections for some axial locations of the SHEV.

Aerospace Science and Engineering - III Aerospace PhD-Days   Materials Research Forum LLC
Materials Research Proceedings 42 (2024) 1-4   https://doi.org/10.21741/9781644903193-1

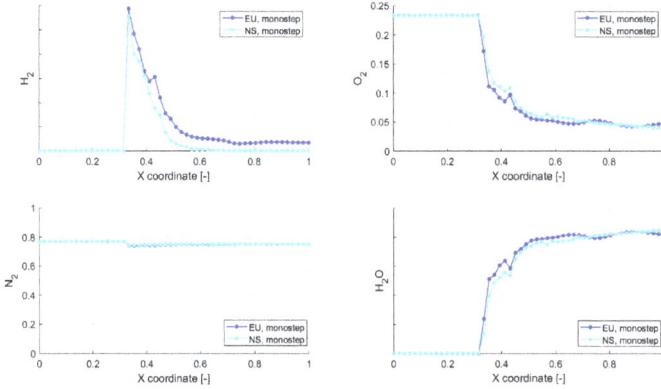

Fig. 4. Comparison of averaged mass fractions profiles for different species along the SHEV internal flowpath at the same altitude. Viscous (NS) and inviscid (EU) effect.

Another comparison between inviscid and viscous effects has been done in order to study the effects on thermodynamic and kinetic dimensionless parameters.

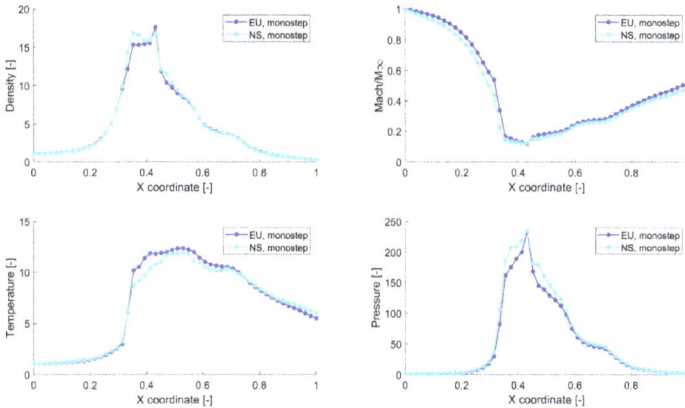

Fig. 5. Comparison of thermodynamic and kinetic averaged parameters profiles for different species along the SHEV internal flowpath at the same altitude. Viscous (NS) and inviscid (EU) effect.

An important parameter that measures the amount of fuel burned compared to that injected at different locations of the SHEV is the combustion efficiency, η_c, defined as:

$$\eta_c = \frac{m_{H_2 \text{ injected}} - m_{H_2}}{m_{H_2 \text{ injected}}} = \frac{m_{H_2 \text{ burned}}}{m_{H_2 \text{ injected}}} \tag{2}$$

As shown in Fig 6, viscosity impacts significantly the engine performance in NS simulations. That seems related to the more properly modelled mixing of the air-fuel mixture that gives an improvement of the combustion efficiency. The air-hydrogen combustion is almost completed in the nozzle, where combustion efficiency reaches a plateau due to frozen condition of the species.

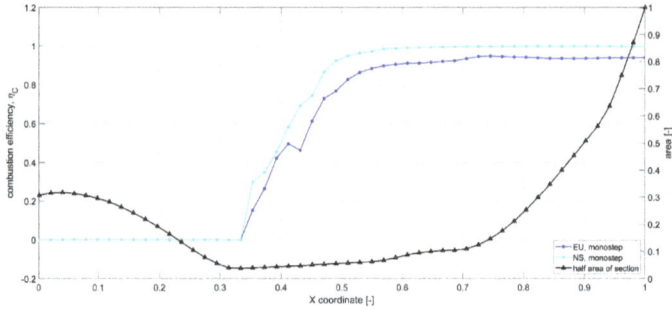

Fig. 6. Comparison of combustion efficiency along internal flow path of SHEV. Viscous (NS) and inviscid (EU) effect.

The project represents a significant research endeavor in the field of scramjet engines, providing a detailed insight into the performance and challenges associated with this advanced propulsion technology in the context of hypersonic flight.

References

[1] Di Benedetto S., Di Donato M.P., Schettino A., Scigliano R., Nebula F., Morani G., Cristillo D., Marini M., Cardone S., Steelant J., Villace V.: The high-speed experimental flight test vehicle of HEXAFLY INT: a multidisciplinary design. CEAS Space Journal. (2021). https://doi.org/10.1007/s12567-020-00341-5

[2] Steelant J. et al., Conceptual Design of the High-Speed Propelled Experimental Flight Test Vehicle HEXAFLY, 20th AIAA International Space Planes and Hypersonic Systems and Technologies Conference, Glasgow, Scotland, AIAA 2015-3539

[3] Viola N., Fusaro R., Saracoglu B., Schram C., Grewe V., Martinez J., Marini M., Hernandez S., Lammers K., Vincent A., Hauglustaine D., Liebhardt B., Linke F., Fureby C.: Main Challenges and Goals of the H2020 STRATOFLY Project. Aerotecnica Missili & Spazio. (2021). 100:95–110, https://doi.org/10.1007/s42496-021-00082-6

[4] Viola N., Roncioni P., Gori O., Fusaro R.: Aerodynamic Characterization of Hypersonic Transportation Systems and Its Impact on Mission Analysis. MDPI-Energies. 16 June 2021. https://doi.org/10.3390/en14123580

[5] Viola, N., Fusaro, R., Saracoglu, B. et al. Main Challenges and Goals of the H2020 STRATOFLY Project. Aerotec. Missili Spaz. 100, 95–110 (2021). https://doi.org/10.1007/s42496-021-00082-6

Aerospace Science and Engineering - IV Aerospace PhD-Days
Materials Research Proceedings 42 (2024) 5-9

Materials Research Forum LLC
https://doi.org/10.21741/9781644903193-2

Selection of best beam theories based on natural frequencies and dynamic response obtained through mode superposition method

Pierluigi Iannotti[1,a]

[1]MUL2 Lab, Department of Mechanical and Aerospace Engineering, Politecnico di Torino, Italy

[a]pierluigi.iannotti@polito.it

Keywords: CUF, Best Theories, Beam Models, AAM, Dynamic

Abstract. Simplified formulations, particularly 1D models, are fundamental for reducing the computational cost typically required by structural analyses. The use of a limited number of nodal degrees of freedom has inevitable implications for the model's capabilities and accuracy. Furthermore, the performance of a reduced formulation is strictly problem-dependent, and the choice of a specific set of primary unknowns must be weighted considering their influence on the accuracy of the results of interest. In this work, a procedure for the selection of the best 1D models to adopt for time-response analyses is investigated. Through the Axiomatic/Asymptotic Method (AAM), the influence of single unknowns is evaluated for a specific structural configuration, which can be described as a combination of aspect ratio, material, geometry, and boundary conditions. The finite element governing equations for every considered set of variables are obtained through the Carrera Unified Formulation (CUF). The main indicator for the quality of a theory is based on the evaluation of a certain number of natural frequencies. Dynamic response analyses are then carried out using the modal superposition method to further asses the performance of the selected best theories.

Introduction

The development of accurate reduced 1D models is a crucial topic in structural mechanics, and it is primarily tied to the need for computational cost reduction. Many efforts were made over the years to improve their capabilities and reduce the gap in accuracy with complete 3D formulations, resulting in a wide variety of approaches [1-4]. Among them, the adoption of higher-order polynomial expansions to describe the displacement field above the cross-section proved remarkably successful, allowing for a proper but still efficient modeling of more complex mechanical behaviors [5].

However, the performances of higher-order theories (HOT) are strictly problem-dependent, and a further increase in the order of the expansion would result in undesired growth of computational demand. These aspects highlight the need for the definition of a theory selection approach able to optimize the number of variables to include in a model, identifying the most influential ones to opt for the derivation of accurate results concerning a specific application.

In this direction, a powerful tool to guide the modeling process is the Axiomatic/Asymptotic Method [6-8]. By directly comparing the results stemming from different combinations of expansion terms and a reference solution, the AAM can inform about the achievable accuracy given the level of complexity and related computational cost of a specific theory. AAM-like procedures require many results to be compared. These can be conveniently obtained in the framework of the Carrera Unified Formulation [9], which provides a generalized methodology for the implementation of structural theories of any type and order, independently from the considered structural problem.

Aerospace Science and Engineering - III Aerospace PhD-Days   Materials Research Forum LLC
Materials Research Proceedings 42 (2024) 5-9   https://doi.org/10.21741/9781644903193-2

Finite Element Formulation

For beam-like structures, the finite element models are built using the reference system presented in Fig. 1, with the x-z plane laying on the cross-section, referred to as Ω.

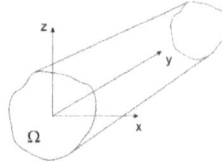

Figure 1 - Reference coordinate system for beam model

The displacement field can be expressed as:

$$\mathbf{u}(x, y, z) = \{u_x, u_y, u_z\}^T \tag{1}$$

In the framework of the Carrera Unified Formulation, 1D models can be refined to provide a better description of the cross-sectional mechanical behavior. The displacement field on the cross-section at a specific y coordinate can be modeled by introducing expansion functions F_τ and F_s:

$$\mathbf{u}(x, y, z) = F_\tau(x, z)\mathbf{u}_\tau, \qquad \delta\mathbf{u}(x, y, z) = F_s(x, z)\delta\mathbf{u}_s \tag{2}$$

$$\tau, s = 1, \dots, M$$

The Einstein notation acts on subscripts τ and s. M denotes the order or number of terms of the expansion, δ is the variational operator used to express the virtual variations. Here, $\mathbf{u}_\tau$ represents the generalized displacement variables involved in the expansion functions ($\delta\mathbf{u}_s$ being their virtual variations). In this work, only expansion functions based on Taylor polynomials were used.

The Finite Element discretization over the beam axis can be obtained by introducing the shape functions $N_{i,j}$. The previous equations thus become:

$$\mathbf{u}(x, y, z) = N_i(y)F_\tau(x, z)\mathbf{q}_{\tau i}, \qquad \delta\mathbf{u}(x, y, z) = N_j(y)F_s(x, z)\delta\mathbf{q}_{sj} \tag{3}$$

$$\tau, s = 1, \dots, M \qquad\qquad i, j = 1, \dots, N_n$$

where N_n is the number of FE nodes, $\mathbf{q}_{\tau i}$ and $\delta\mathbf{q}_{sj}$ are the vectors of nodal unknown variables. For a full beam model of order M=2, the first of Eqs. 3 can be written in extended form as:

$$u_x = u_{x1} + xu_{x2} + zu_{x3} + x^2u_{x4} + xzu_{x5} + z^2u_{x6}$$

$$u_y = u_{y1} + xu_{y2} + zu_{y3} + x^2u_{y4} + xzu_{y5} + z^2u_{y6} \tag{4}$$

$$u_z = u_{z1} + xu_{z2} + zu_{z3} + x^2u_{z4} + xzu_{z5} + z^2u_{z6}$$

with six nodal displacement variables for each component, resulting in a total of eighteen.

The Principle of Virtual Displacements (PVD) is used to derive the governing differential equations. The complete formulation of the dynamic problem can be obtained through a hierarchical and generalized assembly procedure [9] over all nodes, resulting in:

$$\mathbf{M}\ddot{\mathbf{q}}(t) + \mathbf{K}\mathbf{q}(t) = \mathbf{P}(t) \tag{5}$$

The undamped dynamic problem can be solved using the mode superposition method [10, 11], which is more computationally efficient than direct integration schemes such as the Newmark method. The mode superposition method involves the transformation of Eq. 5 into modal coordinates. The finite element nodal point displacements are then obtained by superposition of the response in each mode.

Axiomatic/Asymptotic Method
Techniques like the Axiomatic/Asymptotic Method can be used to assess the impact of individual expansion terms on the model accuracy. By directly comparing the solutions obtained from all possible combinations of generalized unknown variables, the AAM can highlight the most influential terms given a particular structural configuration.

The preliminary step of an Axiomatic/Asymptotic procedure is the selection of a reference solution, typically provided by a full high-order expansion or 3D formulation.

The insights offered by the AAM can be conveniently summarized through the Best Theory Diagram, a graphical representation of the relationship between the number of adopted unknowns and achievable accuracy.

Preliminary Results
The proposed methodology is here introduced considering a simply-supported multi-bay box beam [12]. The sides of the section are b = 0.38 m and h = 0.14 m, with thickness of the wall t = 0.02 m and length L = 10b. An isotropic material was considered, with E=75 GPa, v=0.33, and density ρ = 2700 kg/m^3. The beam was discretized using 10 B4 elements, and only theories up to the fourth order were considered. The full fourth-order expansion (E4) was also used as the reference model. Considering that the E4 model has 45 nodal unknowns, 2^{45} theories should have been compared.

To reduce the required results, the corresponding expansion terms in all three displacement components were activated/deactivated together, thus requiring only 2^{15} total computations. Furthermore, the constant and linear terms of the expansion were always kept active, for a final total of 2^{12} compared theories.

The selection of the best theories was performed using the average error over the first thirty natural frequencies:

$$\%E_{AVG} = \frac{1}{30} \cdot \sum_{i=1}^{30} 100 \cdot \frac{\left|f_i - f_i^{E4}\right|}{f_i^{E4}} \tag{6}$$

Table 1 provides the resulting best-performing models. A black triangle signals an active term of expansion. Follows the corresponding Best Theory Diagram in Fig. 2.

Each determined best theory can then be tested through the comparison of the dynamic response to a time-varying load. For the presented case, two out-of-phase sinusoidal loads were applied at points P_1 (b/2-t/2, L/2, h/2) and P_2 (-b/2+t/2, L/2, -h/2), both with an amplitude F_0 = 10000 N and ω = 5 rad/s. The vertical displacements over time at P_1 obtained with some of the found Best Theories are compared in Fig. 3. The provided results demonstrate the weight that different expansion terms have on the solution accuracy for this specific structural case and the computational advantage that model optimization can guarantee.

Materials Research Forum LLC
https://doi.org/10.21741/9781644903193-2

Table 1 - Best Theories for simply-supported multi-bay box beam based on the average percentage error over the first thirty natural frequencies.

Nodal DOFs	const.	x	z	x^2	xz	z^2	x^3	x^2z	xz^2	z^3	x^4	x^3z	x^2z^2	xz^3	z^4	$\%E_{AVG}$
45	▲	▲	▲	▲	▲	▲	▲	▲	▲	▲	▲	▲	▲	▲	▲	0.0000
42	▲	▲	▲	▲	▲	▲	▲	▲	▲	▲	▲	▲	△	▲	▲	0.0263
39	▲	▲	▲	▲	▲	▲	▲	▲	▲	▲	△	▲	△	▲	▲	0.0329
36	▲	▲	▲	▲	▲	▲	▲	▲	▲	▲	△	▲	△	▲	△	0.0803
33	▲	▲	▲	▲	▲	▲	▲	▲	△	▲	△	▲	△	▲	△	0.1478
30	▲	▲	▲	▲	▲	▲	▲	△	△	▲	△	▲	△	▲	△	0.4045
27	▲	▲	▲	▲	▲	▲	▲	△	△	▲	△	△	△	▲	△	0.8190
24	▲	▲	▲	▲	▲	▲	△	△	△	▲	△	△	△	▲	△	1.3431
21	▲	▲	▲	▲	▲	▲	△	△	△	▲	△	△	△	△	△	2.8689
18	▲	▲	▲	▲	▲	△	△	△	△	▲	△	△	△	△	△	4.7301
15	▲	▲	▲	▲	▲	△	△	△	△	△	△	△	△	△	△	7.1115
12	▲	▲	▲	△	▲	△	△	△	△	△	△	△	△	△	△	10.6242
9	▲	▲	▲	△	△	△	△	△	△	△	△	△	△	△	△	22.9546

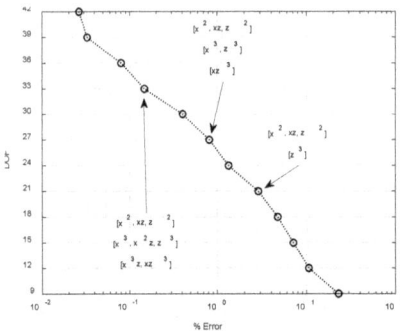

Figure 2 – Best Theory Diagram for simply-supported multi-bay box beam based on the average percentage error over the first thirty natural frequencies.

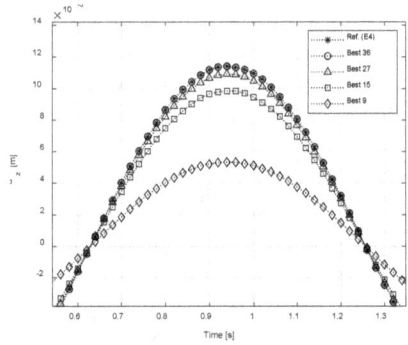

Figure 3 – Time response for simply - supported multi-bay box beam, u_z evaluated in P_1.

Conclusions

This work investigates a methodology for selecting the best beam models for a specific structural configuration.

The Axiomatic/Asymptotic Method is employed to derive the best theories due to the marked problem dependence of the models' accuracy.

Free-vibration and time response analyses are considered, and a performance indicator based on the quality of the estimated natural frequencies is adopted. The study highlights the influence of specific expansion terms on the provided solutions' accuracy and the selection criterion's influence on the outcome of the AAM.

References

[1] E. Carrera, A. Pagani, M. Petrolo, E. Zappino, Recent developments on refined theories for beams with applications, Mechanical Engineering Reviews 2.2 (2015) 14-00298. https://doi.org/10.1299/mer.14-00298

[2] J.N. Reddy, On locking-free shear deformable beam finite elements, Computational Methods in Applied Mechanics and Engineering, 149 (1997) 113–132. https://doi.org/10.1016/S0045-7825(97)00075-3

[3] K. Kapania, S. Raciti, Recent advances in analysis of laminated beams and plates, part I: shear effects and buckling, AIAA Journal 27.7 (1989) 923–935. https://doi.org/10.2514/3.10202

[4] K. Kapania, S. Raciti, Recent advances in analysis of laminated beams and plates, part II: vibrations and wave propagation, AIAA Journal 27.7 (1989) 935–946. https://doi.org/10.2514/3.59909

[5] E. Carrera, M. Petrolo, On the effectiveness of higher-order terms in refined beam theories, Journal of Applied Mechanics 78.2 (2011) 021013. https://doi.org/10.1115/1.4002207

[6] E. Carrera, F. Miglioretti, M. Petrolo, Computations and evaluations of higher-order theories for free-vibration analysis of beams, Journal of Sound and Vibration 331 (2012) 4269-4284. https://doi.org/10.1016/j.jsv.2012.04.017

[7] S. Candiotti, J. L. Mantari, J. Yarasca, M. Petrolo, E. Carrera, An axiomatic/asymptotic evaluation of best theories for isotropic metallic and functionally graded plates employing non-polynomic functions, Aerospace Science and Technology 68 (2017) 179-192. https://doi.org/10.1016/j.ast.2017.05.003

[8] M. Petrolo, P. Iannotti, Best Theory Diagrams for laminated composite shells based on failure indexes, Aerotecnica Missili & Spazio 102 (2023), 199-218. https://doi.org/10.1007/s42496-023-00158-5

[9] E. Carrera, M. Cinefra, M. Petrolo, E. Zappino, Finite element analysis of structures through unified formulation, Wiley & Sons Ltd, 2014.

[10] K.J. Bathe, Finite element procedures, Klaus-Jurgen Bathe, 2006.

[11] O. E. Hansteen, K. Bell, On the accuracy of mode superposition analysis in structural dynamics, Earthquake Engineering & Structural Dynamics, 7.5 (1979) 405-411. https://doi.org/10.1002/eqe.4290070502

[12] M. Dan, A. Pagani, E. Carrera, Free-vibration analysis of simply-supported beams with solid and thin-walled cross-sections using higher-order theories based on displacement variables, Thin-Walled Structures 98 (2016), 478-495. https://doi.org/10.1016/j.tws.2015.10.012

Aerospace Science and Engineering - IV Aerospace PhD-Days   Materials Research Forum LLC
Materials Research Proceedings 42 (2024) 10-13   https://doi.org/10.21741/9781644903193-3

Design, realization and testing of an azimuth-stabilized modular stratospheric platform

Irene Marsili[1,a,b]

[1]Department of Civil and Industrial Engineering – Aerospace Division, University of Pisa, Via Girolamo Caruso 8, 56122, Pisa, PI, Italy

[a]irene.marsili@phd.unipi.it, [b]irene.marsili@unitn.it

Keywords: Attitude Control, Azimuth, High Altitude Platform, Reaction Wheel, Sounding Balloon, Stratosphere

Abstract. In recent years, increasing investments from space industry are directed towards single or constellations of small satellites orbiting in LEO, which have significantly low cost with respect to other types of space missions, but require many years of development and large sums on an absolute scale. The development of new facilities that can cut costs during this phase is therefore of primary interest. One of the main cost items is the development and testing of space hardware, which shall be exposed to high-vacuum and high-level-radiation environment, that is both extremely difficult and expensive to recreate in laboratory. The cheapest option is the use of near-space platforms lifted by small-scale sounding balloons; however, they are not commonly employed because state-of-the-art gondolas for such balloons do not offer a stabilized frame. The work presented in this paper shows a successful attempt at designing and realizing a low-cost yaw-stabilized stratospheric platform, able of self-orienting at set azimuth direction. Active attitude stabilization was achieved via the use of a single reaction wheel placed along the main vertical axis of the platform, controlled by a PI software, and was accompanied by passive attitude stabilization techniques. Tests prove that a stabilized position can be achieved within ±5° of the selected pointing direction, which is preset or directly inputted by the user during platform operations. In this paper, a first application of the platform to host a payload of four solar cells is presented, but particular emphasis was given to modular design, so that the integration of a different payload would require the minimum amount of re-design.

Introduction

Modern scientific ballooning is targeting the development and implementation of high-altitude pseudo satellites (HAPS) flying in the near-space region, that extends from 20 to 100 km in altitude. Applications for HAPS, other than meteorology, range from remote sensing to telecommunication such as, but not limited to telescopes, radio occultation and regional monitoring [1, 2, 3]. Sounding balloons are the small-scale category of stratospheric balloon and have the advantage of being at least two orders of magnitude cheaper than the zero-pressure and super-pressure large-scale counterparts, at the expense of a reduction in payload capability and time of flight, which is usually no longer than 3,5 hours [1], yet reaching the same maximum altitude of 30-35 km. However, many studies to obtain an extension in flight time via altitude stabilization have been made recently, therefore a larger number of users could be interested in this kind of balloons in the future. Nevertheless, many HAPS applications require active attitude stabilization, which is a state-of-the-art technology on platform for large-scale balloons, but not a feature currently offered by platforms for sounding balloon. Active attitude stabilization on small-scale gondolas can be implemented using different strategies, such as reaction wheels [4], motorized pivots [5], cold gas thrusters [6] and propellers [7], but no standardized practices are present in literature. Moreover, since at state-of-the-art a custom gondola tailored to the specific instrument is developed for each launch and/or has not an active attitude stabilization system, no proper HAPS in literature can be accounted for. The work presented in this paper addresses this issue and describes the development, realization and testing of a small-scale reaction-wheel-azimuth-stabilized HAPS which can be suited for different instruments, when properly joined with a custom interface. As a first application example, four solar cells for space have been selected as payload.

Driving Requirements and Platform Design

The purpose of this work was the production of an azimuth-stabilized stratospheric platform suited for sounding balloons and able to accommodate scientific instrumentation of different nature. Starting from this statement, different requirements were defined.

Structural Requirements.
- *Total mass < 4kg* - to be compliant with the ICAO rules;
- *Modular design* - to accommodate payload of different nature, avoiding radical redesign.

Attitude Control System Requirements.
- *Azimuth stabilization* - i.e. around vertical axis;
- *Pointing capability* - stabilization shall be of $\pm 5°$ around the selected azimuth;
- *Decoupling capability* - techniques to promote decoupling from roll-pitch motion and launch train motion shall be taken into consideration during its design.

Functional Requirements.
- *Telemetry and data accessibility* - telemetry must be accessed during platform operations, via a wireless connection, and produced data must be stored on board;
- *Payload accommodation* - shall accommodate 4 solar cells and their electronics;
- *Monitoring capability* - the platform shall be provided with imaging system that shall take pictures of the attitude control system and the payload to assess their correct functioning.

Starting from requirements, all tasks that the platform should be able to perform were defined and for each task, the suitable instrumentation was selected. Following the preliminary evaluation, single components were organized in different subsystems, which are: imaging, attitude control and power supply. The subsystems were subsequently integrated and structural design was conducted to obtain the most compact and lightweight platform as possible, which was 3D modeled and then 3D printed using PLA filament. The basic unit of the main body is of cubic shape, with a 15cm edge. To implement a modular design, the interior of the cube was divided with planes and the reaction wheel was placed on the top part of the platform. Interface for the payload was designed so that they could be exposed to sunlight in the most effective way, and it is placed below the main structure. Payload electronics are accommodated inside the platform, on the bottom plane.

Testing and Results

Indoor experiments were conducted: the platform was suspended to simulate the actual operational condition and a fan was directed towards the platform to recreate disturbances generated by winds' action.

Performance Assessments. Performances were assessed measuring the platforms efficiency in matching the assigned azimuthal direction and maintaining the position, with minor oscillations around the equilibrium point. After the power on and the attitude sensor calibration, starting form rest, the platform oriented towards the magnetic North and kept itself around said position. Then, two different target azimuth commands were sent in sequence. After the third equilibrium point was reached, the test was ended. The total duration of the test was fixed not to be less than 4 minutes and 30 s were awaited after the stabilization before sending the new command. As an example, an individual test, called test A will be analyzed. Fig. 1 depicts the graphs of the azimuth

Aerospace Science and Engineering - III Aerospace PhD-Days   Materials Research Forum LLC
Materials Research Proceedings 42 (2024) 10-13   https://doi.org/10.21741/9781644903193-3

and unwrapped azimuth as a function of time. From the upper graph, the presence of three distinct stabilization segments around 0°, 60° and 270° is evident. Rotations from a target azimuth to the subsequent one are managed performing the minimum angular displacement either in clockwise or counterclockwise rotation. A slight tilting of the actual azimuth stabilization angle with respect to the target one is noticeable, due to the non-linear motor stall of the brushed motor. Nevertheless, the recorded medium of absolute tilting with respect to nominal heading is of 1.72° and the maximum deviation around the actual stabilization value is of 2.45°, in compliance with pointing requirements.

Operational Limits. The reaction wheel saturates at known angular velocity thresholds called ω_t, known from controller gain. A test was conducted to assess the behaviour in case of saturation: after the power on of attitude control system and the attitude sensor calibration, the platform was kept at a stable position and quickly rotated in clockwise direction, so to induce saturation. A sufficient time was then awaited so that the platform could stop its spin and position itself around the target 0° azimuth. Once the platform stabilized around the target azimuth, the procedure was repeated once again in clockwise direction and twice in counterclockwise direction. The overall duration of a single test was fixed not to be less than 4 minutes. As an example, an individual test will be presented, referred to as test C. Upper graph depicted in Fig. 2 shows a peak in correspondence of the instance where the rotation is applied. Middle graph indicates the PWM value, which is in the interval [−255, 255]: *saturation segment* - corresponds to the flat part of the graph; *rotation segment* - it corresponds to the highly dense part of the graph; *stabilization segment* - has an oscillatory pattern. Comparing upper and lower graph, it is evident that during the saturation segment the platform is subjected to full rotations, that progressively become slower as

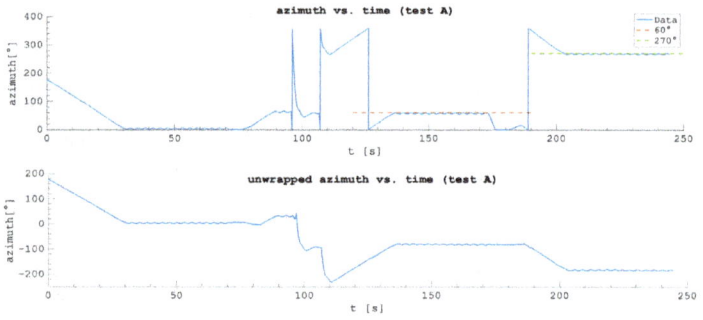

Figure 1. Azimuth (up) and correspondent unwrapped azimuth (down) recorded during test A

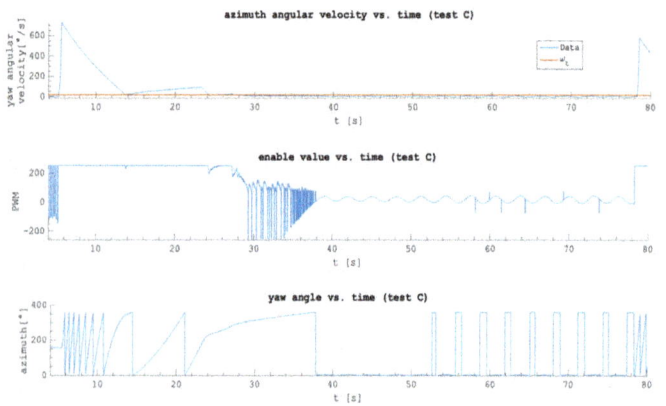

Figure 2. From top to bottom graphs: azimuth angular velocity, PWM, azimuth angle for the first clockwise disturbance in test C

12

the platform decelerates. In the final part of the saturation segment and throughout the rotation segment, the platform is rotated in a controlled approach towards the target azimuth, decelerating a gradually until the 0° azimuth is reached. When the stabilization segment is reached, PWM values are very far from saturation. It is therefore evident that reaction wheel desaturates spontaneously and efficiently, independently from its starting angular speed, due to the continuous exchange of momentum between the platform and the surrounding air.

Conclusion

The successful development of a modular stratospheric platform was demonstrated. Tests have shown that stabilization is achieved within the limits of ±5°. Despite the great results obtained, the platform has only been tested indoor and few open issues remain to be solved. The platform will be subjected in the future to: the integration of a low-power-consumption long-range telecommunication system; mathematical modeling and direct testing, with launch campaigns, to analyze variations of desaturation effect offered by air and torque disturbances due to wind action with altitude; the replacement of the motor with a brushless one, to reduce tilting in stabilization angle and oscillations; the integration of solar cells to improve its operational life, which is of particular interest for altitude-stabilized balloon application.

Acknowledgments

This extended abstract was produced while attending the PhD program in PhD in Space Science and Technology at the University of Trento, Cycle XXXIX, with the support of a scholarship financed by the Ministerial Decree no. 118 of 2nd march 2023, based on the NRRP - funded by the European Union - NextGenerationEU - Mission 4 "Education and Research", Component 1 "Enhancement of the offer of educational services: from nurseries to universities" - Investment 4.1 "Extension of the number of research doctorates and innovative doctorates for public administration and cultural heritage".

References

[1] M. Gemignani and S. Marcuccio, "Dynamic characterization of a high-altitude balloon during a flight campaign for the detection of ISM radio background in the stratosphere," *Aerospace*, vol. 8, no. 1, pp. 1–20, Jan. 2021. https://doi.org/10.3390/aerospace8010021

[2] M. Martorella and E. Aboutanios, "BalSAR: A Stratospheric Balloon-Borne SAR System," in *Advanced Technologies for Security Applications*, Dordrecht: Springer Netherlands, 2020, pp. 283–294. https://doi.org/10.1007/978-94-024-2021-0_25

[3] M. Albano *et al.*, "Hemera: The European Community for Advanced Research in the Stratosphere," *Aerotecnica Missili & Spazio*, vol. 102, no. 4, pp. 377–383, Dec. 2023. https://doi.org/10.1007/s42496-023-00171-8

[4] N. Khoi Tran, D. Evan Zlotnik, and J. Richard Forbes, "Design of an Attitude Control System for a High-Altitude Balloon Payload," in *Academic High Altitude Conference* , Upland, Indiana, Jun. 2013. https://doi.org/https://doi.org//ahac.8149

[5] T. Nakano, "Design of Attitude Control System for Stratospheric Balloon Gondolas by Sliding Mode Control," in *IOP Conference Series: Materials Science and Engineering*, Institute of Physics Publishing, Apr. 2019. https://doi.org/0.1088/1757-899X/501/1/012010

[6] "High Altitude Visual Orientation Control (HAVOC)." Accessed: May 30, 2023. [Online]. Available: https://space.uah.edu/programs/balloonsat/havoc#h.qbzuaslzx4xi

[7] S. L. Kushal and et al., "Airframe Stabilization and Control System for Near Space Balloon Payloads," in *AIAA AVIATION Forum and Exposition 2021*, AIAA Inc., 2021. https://doi.org/10.2514/6.2021-2692

Aerospace Science and Engineering - IV Aerospace PhD-Days   Materials Research Forum LLC
Materials Research Proceedings 42 (2024) 14-17   https://doi.org/10.21741/9781644903193-4

Compressibility and wall-cooling effects on high-speed turbulent boundary layers

Michele Cogo[1] *, Mauro Chinappi[2], Matteo Bernardini[3], and Francesco Picano[1,4]

[1] Centro di Ateneo di Studi e Attività Spaziali 'Giuseppe Colombo' (CISAS), Università degli Studi di Padova, via Venezia 15, 35131 Padua, Italy

[2]Department of Industrial Engineering, University of Rome Tor Vergata, via del Politecnico 1, 00133 Rome, Italy

[3]Department of Mechanical and Aerospace Engineering, Sapienza University of Rome, via Eudossiana 18, 00184 Rome, Italy

[4]Department of Industrial Engineering, Università degli Studi di Padova, via Venezia 1, 35131 Padua, Italy

*michele.cogo.1@phd.unipd.it

Keywords: Compressible Boundary Layers, Turbulent Boundary Layers

Abstract. Flight systems operating at high speeds are frequently enveloped by turbulent, hot, highly compressible flow experiencing various regimes in terms of Mach numbers and surface temperature. These factors are dominant in the proximity of the surface, the boundary layer, posing technical difficulties in predicting the mechanical and thermal loads. In this work, we analyze a database of Direct numerical Simulations of turbulent boundary layers for Mach numbers 2, 4 and 6 and four different wall temperatures (from adiabatic to cold wall). We discuss the choice of the diabatic parameter to define the degree of wall-cooling across different Mach numbers, observing its influence on instantaneous fields and first-second order velocity and temperature statistics.

Introduction and database setup

High speed turbulent boundary layers represent a fundamental case study to determine the aerodynamic heating and drag acting on supersonic and hypersonic vehicles.

As the velocity of the freestream flow becomes several times larger than the speed of sound, i.e. the Mach number increases, compressibility influences the mean and fluctuating fields of thermodynamic quantities, which are coupled to the momentum trough density. When this high-speed flow is brought to rest by the friction with the wall, it is important to consider the amount of kinetic energy that can be converted to thermal, which can be quantified looking at the recovery temperature $T_r = T_\infty(1 + r(\gamma - 1)/2M_\infty^2$ with $r = \text{Pr}^{1/3}$ the recovery factor, since the process is non-isentropic.

In the case that the wall only exchanges heat with the flow for long times, reaching the recovery temperature, the temperature profile would reach a zero-gradient slope, settling in an adiabatic condition.

However, in practical cases, the vehicle surface is usually expected to be at lower temperatures, i.e. $T_w/T_r < 1$, generating large heat fluxes to the wall (cold wall). This affects the flow dynamics in conjunction with the Mach number, enriching the physical effects that need consideration in the formulation of theoretical relations and models.

The presence of a cold wall leads to an alteration in the mean temperature gradient's sign near the wall. This change influences the generation of temperature fluctuations, potentially causing a significant deviation from the similarity between velocity and temperature fields, a building block of theory of compressible flows [6]. This loss of similarity is evident in instantaneous snapshots of turbulent structures [3]. However, this study also noted that the same T_w/T_r condition at

Aerospace Science and Engineering - III Aerospace PhD-Days
Materials Research Proceedings 42 (2024) 14-17

Materials Research Forum LLC
https://doi.org/10.21741/9781644903193-4

different Mach numbers resulted in very different near-wall dynamics for temperature fluctuations, being more similar to velocity at high Mach numbers. Recently, other definitions of the wall temperature condition have been proposed, such as the diabatic parameter $\Theta = (T_w - T_\infty)/(T_r - T_\infty)$ which are capable to account for the indirect effect of Mach number on the wall temperature condition. In fact, this parameter only takes in to account the ΔT that is recovered when the flow is brought at rest, being the only one responsible for kinetic-internal energy exchanges.

In this work, we aim to leverage the diabatic parameter Θ to analyse the interplay of the Mach number and wall temperature condition, discussing their respective imprints on turbulence statistics.

To pursue this objective, an extensive Direct Numerical Simulation (DNS) database consisting of 12 simulations of zero-pressure-gradient turbulent boundary layers has been computed fixing the friction Reynolds number, while spanning three Mach numbers $M_\infty = [2, 4, 6]$ and four diabatic parameters $\Theta = \{0.25, 0.5, 0.75, 1.0\}$, going from extremely cold walls, $\Theta = 0.25$, to adiabatic case, $\Theta = 1$.

Information concerning the domain size, grid resolution and boundary layer properties at selected stations (where $Re_\tau = 443$) are available at reference [4].

Here, M_∞ is the free-stream Mach number and Re_τ is the friction Reynolds number, defined as the ratio between the boundary layer thickness δ_{99} and the viscous length scale $\delta_v = v_w/u_\tau$, where $u_\tau = \sqrt{\tau_w/\rho_w}$ is the friction velocity, τ_w is the mean wall shear stress, and v_w is the kinematic viscosity at the wall. Moreover, $y^* = y/\delta_{v,SL}$ is the semilocal scaled wall-normal coordinate, with $\delta_{v,SL} = v/\sqrt{\tau_w/\rho}$.

The Navier–Stokes equations are solved using STREAmS [2], an open-source numerical solver for compressible flows oriented to modern HPC platforms using MPI parallelization and supporting multi-GPU architectures under the CUDA paradigm. In shock-free regions, the convective terms are evaluated using high-order energy preserving schemes, while a shock capturing formulation is activated with a Ducros sensor. The diffusive terms are discretized using a locally conservative scheme expanded to standard Laplacian formulation.

Results

We first consider the instantaneous field in the form of wall-parallel slices of velocity and temperature fluctuations, Figure 1. The slices are selected at a wall-normal location $y^* = 10$, which is approximately coincident with the peak of streamwise velocity fluctuations in the buffer layer [4]. Here, we selected the two extreme cases of the diabatic parameter $\Theta = \{0.25, 1\}$ and Mach numbers $M_\infty = \{2, 6\}$.

As expected, all cases show the presence of near-wall streaks for velocity fluctuations with similar intensities, a sign of the near-wall self-sustaining cycle of turbulence.

However, the picture is clearly different for temperature fluctuations. While adiabatic cases maintain a streaky pattern very similar to the velocity counterpart, showing a clear temperature-velocity coupling, cold cases ($\Theta = 0.25$) show highly damped fluctuations with an absence of elongated streaks.

These qualitative results preliminarily show two aspects: the first is that cold walls reduce the temperature-velocity coupling in the buffer layer, highly damping temperature fluctuations. The second aspect is that fixing the diabatic parameter yields the same general behavior due to the effect of wall-cooling at different Mach numbers.

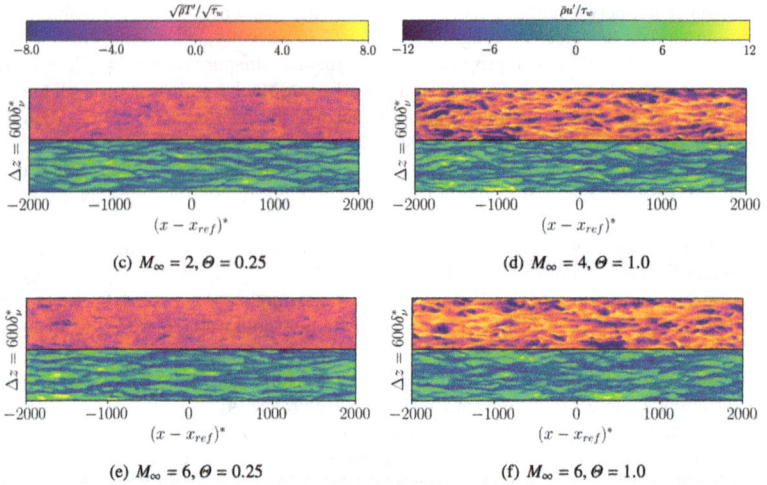

Figure 1: Temperature (top) and streamwise velocity (bottom) fluctuations in wall-parallel slices (x-z plane) selected at $y^* \approx 10$. Here, only the cases at Mach numbers 2 and 4 are shown with the respective extremes in $\Theta = \{0.25, 1\}$.

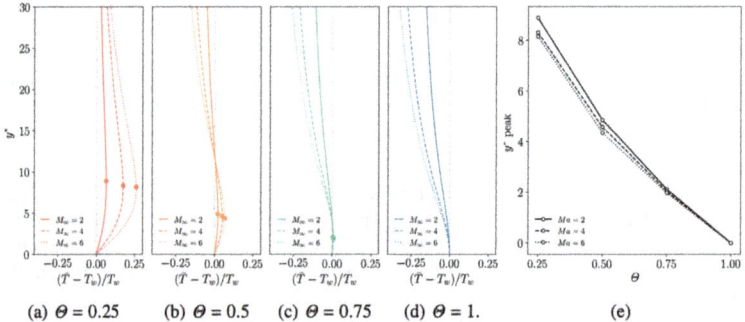

Figure 2: Panels (a-b-c-d): Mean temperature profiles scaled with T_w and relative peaks as function of the wall-normal coordinate y^*. Panel (e): Wall-normal location of mean temperature peaks as function of the diabatic parameter Θ.

An insight on the underlying physics regulating the decoupling between temperature and velocity fluctuations can be obtained by looking at wall-normal mean temperature profiles, Figure 2. As the Mach number increases, the recovery temperature at which the flow would like to settle at the wall increases (quadratically with M_∞). However, a cold wall condition ($T_w < T_r$) forces the temperature profile to slant towards colder temperature, resulting in a local peak, evident in Figure 2(a). Here, it is also evident that the prominence of this peak is depended on the Mach number, and gives rise to the aerodynamic heating, generating a net heat flux from the flow to the solid boundary. Local temperature peaks are marked in figures 2(a) to 2(d) with dots for all diabatic parameters. An increase in the Mach number generates more intense gradients and higher peak temperatures for non-adiabatic cases, enhancing aerodynamic heating. However, the wall-normal

Aerospace Science and Engineering - III Aerospace PhD-Days   Materials Research Forum LLC
Materials Research Proceedings 42 (2024) 14-17   https://doi.org/10.21741/9781644903193-4

location of the peaks is mainly dependent by the change Θ, and only weakly dependent on the Mach number. This is visible in Figure 2(e), which shows a departure from the wall of the peak location as Θ decreases. Given that the mean temperature gradient is directly connected to thermal production, $P_T = -\bar{\rho}\,\widehat{v''T''}\,\partial\bar{T}/\partial y$ (refer to [5]), and that for the cases ($\Theta = 0.25$) $\partial\bar{T}/\partial y \approx 0$ at $y^* \approx 10$, when can see why temperature fluctuations are highly damped in this region. As a result, we observe an absence of near-wall streaks and a reduction of their overall intensity, as visible in Figure 1.

Conclusions

In this study, we presented a database of 12 DNS highlighting the effect of the Mach number and wall-cooling on zero-pressure-gradient turbulent boundary layers.

The instantaneous flow organization revealed strong differences from temperature and velocity fields as the wall-cooling increased, which were visible at all Mach numbers when fixing the diabatic parameter Θ, a sign of its aptness to yield the same wall-cooling effects across different M_∞.

The temperature-velocity decoupling for cold cases can be explained by looking at the mean temperature profiles, knowing their influence in the thermal production term. We observed that as the wall-cooling increases, there is a formation of a local mean temperature peak that rises in the wall-normal direction. In our coldest case ($\Theta = 0.25$), the location of the peak approaches the buffer layer, highly damping the temperature fluctuations and their correlations with velocity.

References

[1] Zhang, Y. S., Bi, W. T., Hussain, F., & She, Z. S. (2014). A generalized Reynolds analogy for compressible wall-bounded turbulent flows. *Journal of Fluid Mechanics, 739*, 392-420. https://doi.org/10.1017/jfm.2013.620

[2] M. Bernardini, D. Modesti, F. Salvadore, S. Sathyanarayana, G. Della Posta, and S. Pirozzoli. Streams-2.0: Supersonic turbulent accelerated navier-stokes solver version 2.0. *Comput. Phys. Commun.*, 285:108644, 2023. https://doi.org/10.1016/j.cpc.2022.108644

[3] Cogo, M., Salvadore, F., Picano, F., & Bernardini, M. (2022). Direct numerical simulation of supersonic and hypersonic turbulent boundary layers at moderate-high Reynolds numbers and isothermal wall condition. *Journal of Fluid Mechanics, 945*, A30. https://doi.org/10.1017/jfm.2022.574

[4] Cogo M, Baù U, Chinappi M, Bernardini M, Picano F. Assessment of heat transfer and Mach number effects on high-speed turbulent boundary layers. *Journal of Fluid Mechanics.* 2023;974:A10. https://doi.org/10.1017/jfm.2023.791

[5] T. B. Gatski and J. P. Bonnet. *Compressibility, turbulence and high speed flow*. Academic Press, 2013.

[6] Morkovin, M.V. 1962 Effects of compressibility on turbulent flows. Méc. Turbul. 367(380),26

Aerospace Science and Engineering - IV Aerospace PhD-Days   Materials Research Forum LLC
Materials Research Proceedings 42 (2024) 18-21   https://doi.org/10.21741/9781644903193-5

Digital optics and machine learning algorithms for aircraft maintenance

Salvatore Merola[1,a] *, Michele Guida[1,b] and Francesco Marulo[1,c]

[1]Department of Industrial Engineering, University of Naples Federico II, Via Claudio 21, Naples, 80125, Italy

[a]salvatore.merola@unina.it, [b]michele.guida@unina.it, [c]francesco.marulo@unina.it

Keywords: Image Processing, Aircraft Maintenance, Deep Learning, Convolutional Neural Networks, Computer Vision

Abstract. The objective of this study is to present a novel approach for airplane inspection to identify damages on the fuselage. Machine learning algorithms in the aeronautic industry can be used as an instrument for automating the process of inspection and detection, decreasing human error, and increasing productivity and security for operators. An overview of the problems, methods, and recent developments in the field of deep learning algorithms used for general damage detection on aircraft components is provided in this extended abstract. Data were collected using a high-quality acquisition system and the dataset was populated by collecting defect images from 2 typologies of aircraft: a commercial partial full-scale airplane fuselage section in primer paint and a general aviation fuselage white painted, both situated in the laboratory of heavy structures of University of Naples Federico II. The Convolutional Neural Networks (CNNs) and machine learning models were trained on large datasets of annotated images, enabling them to learn complex features associated with different types of damage. Data augmentation techniques are adopted to add diversity to the training data. Transfer learning techniques, which leverage pre-trained models on large-scale image datasets, have also proved to be effective in achieving accurate and robust detection results.

State of the art

As of 2018, the inspection of aircraft is primarily done manually with 80% of the inspection being carried out visually. In the case of aircraft structures, initially, Structural Significant Items (SSI) are identified, whose failure can compromise the structural integrity of the aircraft. Then, categories are assigned based on the type of item identified, which can include fatigue damage, environmental deterioration, and accidental damage. Consequently, the most effective inspection method is chosen, which can be [1, 2, 3]:

- *General visual inspection*: visual examination of the item for damage. This inspection is normally performed at an arm's length and under normal lighting conditions. Sometimes a mirror may be used to ensure visual access to all areas to be inspected.
- *Detailed inspection*: intensive visual examination of the item for damage. An additional light source with appropriate intensity is used. Mirrors and magnifying lenses may be used to make the refurbishment operations more visible.
- *Special Detailed Inspection*: Specific and in-depth examination of the item for damage. Specialized inspection techniques and Non-Destructive Testing (NDT) equipment are extensively used.

Since the early 90's, the field of artificial intelligence has experienced a dramatic reinvigoration. Research and development in this field have been exponentially growing, particularly in the field

Aerospace Science and Engineering - III Aerospace PhD-Days   Materials Research Forum LLC
Materials Research Proceedings 42 (2024) 18-21   https://doi.org/10.21741/9781644903193-5

of deep learning, a broad family of machine learning algorithms within the field of AI. One of the reasons why deep learning has become increasingly popular is that it eliminates hand-picked features, a time-consuming process in classical machine learning. Among the many deep learning architectures, the convolutional neural network [4] is one of the most popular architectures for image classification. To our knowledge, the very first deep CNN application for visual inspection was performed by Zhang et al. [5] for road crack detection on concrete structures by using 6 layers of ConvNet. Since then, similar works using CNN for crack detection have been performed in civil infrastructure such as buildings, pavement, and concrete surfaces [6]. In the Aerospace field, many applications of deep learning are in developing flight mechanical, flight tests, health monitoring of aeronautical structures, and aircraft maintenance [7].

Methodology
The main objective of this topic is to study and identify new methodologies for aircraft maintenance. To do this, many approaches are investigated:

- *Image processing and segmentation*
- *Convolutional neural networks*
- *Computer vision algorithms*

Starting from image processing, it can be possible to investigate some general parameters of digital cameras. The first trade-off phase for the selection of the acquisition system is to set up the digital optics system to be used in the industrial application. In particular, many different systems were compared (in terms of resolution, weight, and size of images) to collect damage examples on a full-scale aircraft section. These images were taken by drones or blimps, which allow for reaching remote areas on the aircraft that are not easily accessible by operators. Convolutional neural networks aim to split images into databases for the training, validation, and test phases. This architecture forms the basis of computer vision algorithms: regions of interest are traced in the images of the database, then the training phase is conducted, and finally, the validation phase is carried out, where the detected types of defects are identified and classified. Computer vision algorithms gave output segmentation in images with the identification and classification of the damages. In the first work, the procedure was to apply the transfer learning technique [8] to a convolutional neural network to identify and classify three types of defects:

- Missing rivet
- Corroded rivet

Transfer learning is a technique in deep learning where a pre-trained model, which has already been learned from a large dataset, is used as a starting point for a new task or a different dataset. Starting from the tracing of regions of interest in the images of a database (constructed with images taken on the fuselage of a civil transport aircraft placed in the laboratory of the heavy structure of the University of Naples Federico II). The next training phase occurs by optimizing two functions: the *accuracy function* and the *loss function*. In particular, the objective is to minimize the loss function to improve the network's performance. Fig. 1 to Fig. 4 represents the results obtained during the testing phase of the CNN. The label "rivetto macchiato" translates to corroded rivet, and the label "rivetto mancante" translates to missing rivet.

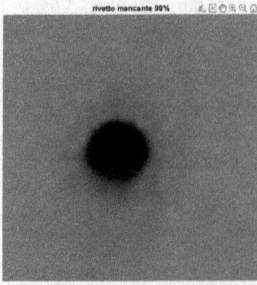

Fig. 1 - Test 1 for CNN

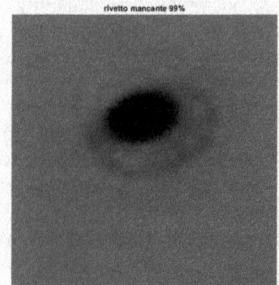

Fig. 2 - Test 2 for CNN

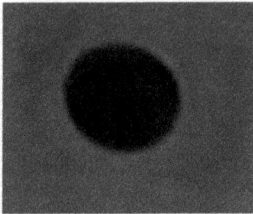

Fig. 3 - Test 3 for CNN

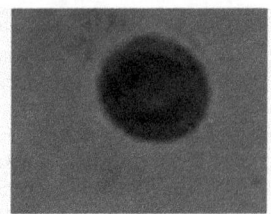

Fig. 4 - Test 4 for CNN

Conclusions

Starting from a trade-off phase for the choice of the acquisition system, an image database was built and used to train a CNN capable of classifying defects on rivets. The trained convolutional neural network shows high accuracy in classifying rivets; in particular, there is a maximum accuracy of 99% for missing rivet class and a maximum accuracy of 87% for corroded rivets class.

Future works

The results obtained from training the CNN previously described have been useful for transitioning to a different approach in image processing: defects were no longer merely classified, but they were also spatially localized during post-process. In future work, an optimization process will be carried out on the network, which will be able to identify and classify a greater number of defects with higher accuracy. Computer vision algorithms aim to replicate human vision capabilities by analyzing and extracting meaningful information from visual inputs. This approach offers operators the possibility to perform near real-time maintenance under conditions of increased safety, both during scheduled maintenance and pre-flight checks.

References

[1] Airbus. ref 51-11-00 Description - DAMAGE CLASSIFICATION. Tech. rep. Airbus Group, 2003.

[2] Airbus. ref 51-11-13 - Description - DAMAGE/DEFECT REPORTING. Tech. rep. Airbus Group, 2003. https://doi.org/10.1016/S1365-6937(03)00036-4

[3] Airbus. ref 51-11-11 Description - Allowable Damage. Tech. rep. Airbus Group, 2003.

[4] Y. LeCun, Y. Bengio, and G. Hinton. "Deep learning". In: nature 521.7553 (2015), pp. 436–444. https://doi.org/10.1038/nature14539

[5] Zhang, Lei, et al. "Road crack detection using deep convolutional neural network." 2016 IEEE international conference on image processing (ICIP). IEEE, 2016. https://doi.org/10.1109/ICIP.2016.7533052

[6] W. R. L. da Silva and D. Schwerz de Lucena. "Concrete cracks detection based on deep learning image classification". In: Proceedings. Vol. 2. 8. MDPI. 2018, p. 489.

[7] M. Rice, L. Li, G.Ying, M. Wan, E.T. Lim, G. Feng, J. Ng, M. N. T. Jin-Li, and V.S. Babu. Automating the Visual Inspection of Aircraft. Proceedings of the Singapore Aerospace Technology and Engineering Conference (SATEC), Singapore. Vol. 7. 2018.

[8] D. Han, Q. Liu, and W. Fan. "A new image classification method using CNN transfer learning and web data augmentation". In: Expert Systems with Applications 95 (2018), pp. 43–56. https://doi.org/10.1016/j.eswa.2017.11.028

Aerospace Science and Engineering - IV Aerospace PhD-Days · Materials Research Forum LLC
Materials Research Proceedings 42 (2024) 22-25 https://doi.org/10.21741/9781644903193-6

Investigation of radiative needle flow dynamics with variable viscosity and thermal conductivity

Niba Kainat[1,a*] and Vincenzo Gulizzi[1,b]

[1] Department of Engineering, Università degli Studi di Palermo, Italy

[a] niba.kainat@unipa.it, [b] vincenzo.gulizzi@unipa.it

Keywords: Radiative Needles, Nonlinear Radiation, Temperature-Dependent Viscosity, Temperature-Dependent Thermal Conductivity

Abstract. The flow pattern formed by the radiative needle moving in a fluid with temperature dependent viscosity and thermal conductivity is investigated in this paper considering the effects of viscous dissipation and nonlinear radiation. Under axial symmetry constraint, the governing equations are converted into a set of non-linear differential equations. Numerical results are presented discussing the influence of Prandtl number, viscosity and thermal conductivity on the heat transfer between the flow and the needle.

Introduction

Boundary layer flows have a pivotal role in aerodynamics affecting phenomena like wing stall, skin friction drag, and heat transfer. In the latter case, which is of scientific and engineering interest for applications as diverse as hypersonic flight, heat exchangers design, or crude oil extraction processes, the boundary layer equations must account for the energy balance equation and the temperature dependence of the viscosity and the thermal conductivity of the fluid. Additionally, at very high temperatures, the magnitude of heat transferred via radiation becomes comparable to that of conduction and convection and should be included as well. Radiative heat transfer also plays a very important role, in general, in flows with high specific enthalpy, in the presence of chemical reactions, dissociation of molecules, ionization of atoms, such as in combustion chambers of chemical rockets [1], in electrical plasma thrusters [2] or during hypersonic atmospheric reentry from extraplanetary space missions or in high enthalpy plasma facilities [3]. However, the effect of radiation on boundary layer flows does not appear fully investigated in the literature, especially when the viscosity and the thermal conductivity are temperature dependent.

The influence of radiation on boundary layer flows over a flat plate was first investigated by Smith [4] upon assuming that convection terms be smaller than conduction terms sufficiently far downstream. Chamkha et al.[5] studied the effect of radiation on the free convection past a vertical plate under the hypothesis of constant viscosity, constant thermal conductivity and linearized heat radiation. Linear radiation effects were also considered by Makinde and Ogulu [6], who however retained the temperature dependence of the viscosity. Pantokratoras [7] investigated linear and nonlinear radiation effects on the natural convection over a vertical plate but considered constant viscosity and thermal conductivity.

The studies above focus on boundary layer flows over flat plates. However, boundary layer solutions have also been found for axial-symmetric problems [8,9]. In the context of radiative heat transfer, Afridi and Qasim [10] recently investigated the effect of nonlinear radiation on the boundary layer flow over a thin needle considering constant viscosity and thermal conductivity.

Based on the literature review above, the present contribution is then intended to study the influence of temperature dependent viscosity and thermal conductivity for a boundary layer flow over a radiative needle. To the best of our knowledge, this problem has not yet been explored.

Problem Analysis

Consider the steady, axisymmetric boundary layer flow over a thin needle assumed as body of revolution, as shown in Fig. 1. The problem is assumed in a cylindrical reference system $O\bar{r}\bar{x}$, whereas $(\bar{r}, \bar{x})$ represent the axial and the radial components, respectively. The needle has radius $R = R(\bar{x})$, and is moving with a velocity $\bar{u}_w$ in a free stream of velocity $\bar{u}_\infty$. Under the boundary layer assumptions, the governing equations for this problem are reduced to

$$\frac{\partial(\bar{r}\bar{u})}{\partial \bar{x}} + \frac{\partial(\bar{r}\bar{v})}{\partial \bar{r}} = 0, \tag{1}$$

$$\bar{u}\frac{\partial \bar{u}}{\partial \bar{x}} + \bar{v}\frac{\partial \bar{u}}{\partial \bar{r}} = \frac{1}{\rho}\frac{\bar{\mu}(\bar{T})}{\bar{r}}\frac{\partial}{\partial \bar{r}}\left(\bar{r}\frac{\partial \bar{u}}{\partial \bar{r}}\right) + \frac{1}{\rho}\frac{\partial \bar{T}}{\partial \bar{r}}\frac{\partial \bar{u}}{\partial \bar{r}}, \tag{2}$$

$$\rho\varsigma_p\left(\bar{u}\frac{\partial \bar{T}}{\partial \bar{x}} + \bar{v}\frac{\partial \bar{T}}{\partial \bar{r}}\right) = \frac{\bar{\kappa}(\bar{T})}{\bar{r}}\frac{\partial}{\partial \bar{r}}\left(\bar{r}\frac{\partial \bar{T}}{\partial \bar{r}}\right) + \left(\frac{\partial \bar{T}}{\partial \bar{r}}\right)^2\frac{\partial \bar{\kappa}(\bar{T})}{\partial \bar{T}} + \bar{\mu}(\bar{T})\left(\frac{\partial \bar{u}}{\partial \bar{r}}\right)^2 + \frac{16\sigma_{SB}}{3a_R}\frac{\partial}{\partial \bar{r}}\left(\bar{T}^3\frac{\partial \bar{T}}{\partial \bar{r}}\right), \tag{3}$$

where $\bar{u}$, $\bar{v}$, and $\bar{T}$ are the axial velocity component, the radial velocity component and the temperature of the fluid, respectively.

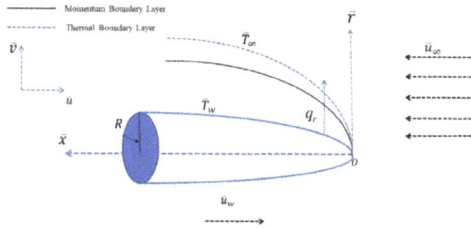

Figure 1. *Flow configuration*

Additionally, ρ is the density, ς_p is the specific heat capacity, $\bar{\mu}$ is the viscosity and $\bar{\kappa}$ is the thermal conductivity of the fluid. As indicated in Eqs. 2 and 3, $\bar{\mu}$ and $\bar{\kappa}$ are functions of temperature; here, the considered functional dependency is

$$\bar{\mu}(\bar{T}) = \frac{\mu_o}{1+\bar{\delta}_\mu(\bar{T}-\bar{T}_\infty)} \text{ and } \bar{\kappa}(\bar{T}) = \kappa_o\left[1 + \bar{\delta}_k\left(\frac{\bar{T}-\bar{T}_\infty}{\bar{T}_w-\bar{T}_\infty}\right)\right], \tag{4}$$

where μ_o, κ_o, $\bar{\delta}_\mu$, and $\bar{\delta}_k$ are material parameters, while $\bar{T}_w$ and $\bar{T}_\infty$ are the needle wall temperature and the free-stream temperature, respectively. The last term in Eq. 3 accounts for the heat exchanged via radiation, where a_R is the Rosseland absorption coefficient and σ_{SB} is the Stefan-Boltzmann constant. The governing equations are closed by the following set of boundary conditions

$$\bar{u} = \bar{u}_w, \bar{v} = 0, \bar{T} = \bar{T}_w \qquad \text{at } \bar{r} = \bar{R}(\bar{x}) \tag{5a}$$

$$\bar{u} \to \bar{u}_\infty, \bar{v} \to 0, \bar{T} \to \bar{T}_\infty \qquad \text{as } \bar{r} \to \infty \tag{5b}$$

Then, considering the following similarity transformations

$$\eta = \frac{\rho(\bar{u}_w+\bar{u}_\infty)\bar{r}^2}{\bar{\mu}\bar{x}}, \bar{\psi} = \frac{\bar{\mu}\bar{x}f(\eta)}{\rho}, \text{ and } \theta(\eta) = \frac{\bar{T}-\bar{T}_\infty}{\bar{T}_w-\bar{T}_\infty}, \tag{6}$$

it is possible to show that the continuity equation is automatically fulfilled, Eq. 2 reduces to

$$(1 + \theta\delta_\mu)[f''(\eta) + \eta f'''(\eta)] - \eta\delta_\mu\theta'f''(\eta) + \frac{1}{2}(1 + \theta\delta_\mu)^2 f(\eta)f''(\eta) = 0, \tag{7}$$

Aerospace Science and Engineering - III Aerospace PhD-Days   Materials Research Forum LLC
Materials Research Proceedings 42 (2024) 22-25   https://doi.org/10.21741/9781644903193-6

while Eq. 3 becomes

$$(1 + \delta_k\theta)(\theta' + \eta\theta'') + \eta(\theta')^2\delta_k + \frac{1}{2}\Pr\theta'f(\eta) + \frac{4\eta\,\Pr\,Ec\,(f''(\eta))^2}{(1+\theta\delta_\mu)} +$$

$$+\frac{4}{3\,N_r}[\theta(\theta_r - 1) + 1]^2\left([\theta(\theta_r - 1) + 1]\left[\eta\theta'' + \frac{\theta'}{2}\right] + 3\eta(\theta')^2(\theta_r - 1)\right) = 0 \qquad (8)$$

In Eq. 8, $\Pr = \varsigma_p\bar{\mu}/\bar{\kappa}$ is the Prandtl number, $Ec = (\bar{u}_w + \bar{u}_\infty)^2/\left(\varsigma_p(\bar{T}_w - \bar{T}_\infty)\right)$ is the Eckert number, $\theta_r = \bar{T}_w/\bar{T}_\infty$ is the heating parameter, and $N_r = a_R\bar{\kappa}/(4\sigma_{SB}\bar{T}_\infty^3)$ is the thermal radiation parameter. Eventually, using Eq. 6, the boundary conditions associated with Eqs. 7 and 8 read

$$f(\bar{a}) = \frac{\bar{a}\varepsilon}{2}, f'(\bar{a}) = \frac{\varepsilon}{2}, \theta(\bar{a}) = 1\text{ at} \qquad\qquad \eta = \bar{a}$$
$$(9a)$$

$$f'(\infty) \to \frac{1-\varepsilon}{2}, \theta(\infty) \to 0\text{ as} \qquad\qquad \eta \to \infty,$$
$$(9b)$$

where $\varepsilon = \bar{u}_w/(\bar{u}_w + \bar{u}_\infty)$ and $\eta = \bar{a}$ identifies the wall surface of the needle.

Results

The set of differential equations given in Eqs. 7 and 8, and their associated boundary conditions given in Eq. 9 are solved numerically. In this study, we use MATLAB built-in bvp5c routine. Some selected numerical results are reported and discussed in this section to investigate the effect of temperature dependent viscosity and thermal conductivity on the heat transfer between the fluid and the radiative needle.

We consider a problem setup identified by $\bar{a} = 0.1$, $\varepsilon = 0.3$, $\theta_r = 2$, and $N_r = 10$, and investigate the effect of the Prandtl number, the Eckert number and the material parameters $\bar{\delta}_k$ and $\bar{\delta}_\mu$ on the heat flux at the needle's wall, i.e. $\theta'(\bar{a})$, and on the local Nusselt number Nu given as

$$\frac{Nu}{\sqrt{Re}} = -2\sqrt{\bar{a}}\left(1 + \frac{4}{3N_r}\theta_r^3\right)\theta'(\bar{a}), \qquad (10)$$

being Re the Reynolds number. The computed results are reported in Table 1, which shows that increasing the Prandtl number leads to an increase of the heat flux; a similar effect is observed if $\bar{\delta}_\mu$ increases. On the other hand, increasing $\bar{\delta}_k$ or Ec lead to a decrease of the heat flux.

Table 1. Effect of the flow parameters on the heat flux between at the needle's wall.

Pr	$\bar{\delta}_k$	$\bar{\delta}_\mu$	Ec	$-\theta'(\bar{a})$	Nu/$\sqrt{Re}$
1.2	0.2	0.3	0.5	1.1120	1.4535
3.0	0.2	0.3	0.5	1.4371	1.8784
7.0	0.2	0.3	0.5	1.8358	2.3995
2.0	0.0	0.3	0.5	1.2884	1.6840
2.0	0.3	0.3	0.5	1.2782	1.6707
2.0	0.6	0.3	0.5	1.2678	1.6571
2.0	0.2	0.0	0.5	1.2649	1.6534
2.0	0.2	0.2	0.5	1.2765	1.6685
2.0	0.2	0.4	0.5	1.2864	1.6815
2.0	0.2	0.3	0.0	1.3553	1.7715
2.0	0.2	0.3	0.5	1.2817	1.6752
2.0	0.2	0.3	1.0	1.2082	1.5792

Conclusions

This paper investigates the impact of temperature dependent viscosity and thermal conductivity on the heat transfer between a fluid and a needle considering nonlinear radiation effects in the energy balance equation. The governing equations of the problems are written in axial symmetric coordinates and transformed into a system of nonlinear ordinary differential equations using a similarity transformation. The system of equations is then solved numerically. The obtained results show that removing the hypothesis of constant viscosity or constant thermal conductivity has a non-negligible effect on the heat flux between the fluid and the needle.

References

[1] Fabiani, M., Gubernari, G., Migliorino, M.T., Bianchi, D., & Nasuti, F. Numerical Simulations of Fuel Shape Change and Swirling Flows in Paraffin/Oxygen Hybrid Rocket Engines. *Aerotecnica Missili & Spazio*, (2023) 102(1), 91-102. https://doi.org/10.1007/s42496-022-00141-6

[2] Majorana, E., Souhair, N., Ponti, F., & Magarotto, M. Development of a plasma chemistry model for helicon plasma thruster analysis. *Aerotecnica Missili & Spazio*, (2021), 100(3), 225-238. https://doi.org/10.1007/s42496-021-00095-1

[3] Esposito, A., & Lappa, M. Perspectives and Recent Progresses on the Simulation of the Entry into the Atmospheres of the Outer Ice Giants. *Aerotecnica Missili & Spazio* (2023), 102(4), 367-376. https://doi.org/10.1007/s42496-023-00167-4

[4] Smith, J.W. Effect of gas radiation in the boundary layer on aerodynamic heat transfer. *Journal of the Aeronautical Sciences*, (1953) 20(8), 579–580. https://doi.org/10.2514/8.2740

[5] Chamkha, A.J., Takhar, H. S., & Soundalgekar, V. M. Radiation effects on free convection flow past a semi-infinite vertical plate with mass transfer. *Chemical Engineering Journal*, (2001) 84(3), 335–342. https://doi.org/10.1016/s1385-8947(00)00378-8

[6] Makinde, O.D., & Ogulu, A. The effect of thermal radiation on the heat and mass transfer flow of a variable viscosity fluid past a vertical porous plate permeated by a transverse magnetic field. *Chemical Engineering Communications* (2008) 195(12), 1575-1584. https://doi.org/10.1080/00986440802115549

[7] Pantokratoras, A. Natural convection along a vertical isothermal plate with linear and nonlinear Rosseland Thermal Radiation. *International Journal of Thermal Sciences*, (2014) 84, 151–157. https://doi.org/10.1016/j.ijthermalsci.2014.05.015

[8] Lee, L.L. Boundary layer over a thin needle. *Physics of fluids*, (1967) 10(4), 820-822. https://doi.org/10.1063/1.1762194

[9] Qasim, M., Riaz, N., Lu, D., & Afridi, M. I. Flow over a needle moving in a stream of dissipative fluid having variable viscosity and thermal conductivity. *Arabian Journal for Science and Engineering*, (2021) 46(8), 7295–7302. https://doi.org/10.1007/s13369-021-05352-w

[10] Afridi, M.I., & Qasim, M. Entropy generation and heat transfer in boundary layer flow over a thin needle moving in a parallel stream in the presence of nonlinear Rosseland radiation. *International Journal of Thermal Sciences*, (2018) 123, 117–128. https://doi.org/10.1016/j.ijthermalsci.2017.09.014

Aerospace Science and Engineering - IV Aerospace PhD-Days   Materials Research Forum LLC
Materials Research Proceedings 42 (2024) 26-30   https://doi.org/10.21741/9781644903193-7

Modal analysis of hyperelastic structures in non-trivial equilibrium states via higher-order plate finite elements

Piero Chiaia[1,a]

[1]Politecnico di Torino, Corso Duca degli Abruzzi, 10129, Turin, Italy

[a]piero.chiaia@polito.it

Keywords: Higher-Order Finite Elements, Plate Models, Hyperelasticity, Compressible Soft Materials, Modal Analysis, Undamped Vibrations

Abstract. The present work proposes a higher-order plate finite element model for the three-dimensional modal analysis of hyperelastic structures. Refined higher-order 2D models are defined in the well-established Carrera Unified Formulation (CUF) framework, coupled with the classical hyperelastic constitutive law modeling based on the strain energy function approach. Matrix forms of governing equations for static nonlinear analysis and modal analysis around nontrivial equilibrium conditions are carried out using the Principle of Virtual Displacements (PVD). The primary investigation of the following study is about the natural frequencies and modal shapes exhibited by hyperelastic soft structures subjected to pre-stress conditions.

Introduction

In recent years, renewed interest in soft materials in diverse fields, including mechanical, aeronautical, robotics engineering, and biological applications, has led to the development of new and efficient computational models for numerical simulations. The enhanced elastic properties of soft hyperelastic structures have attracted many researchers, and the dynamic features of hyperelastic materials have garnered increased attention. These features of hyperelastic structures generally lead to highly nonlinear equilibrium equations that do not allow closed-form solutions for static or dynamic problems. Furthermore, classical structural theories for beams and plates have been proven inadequate when considering large strains and nonlinear constitutive law. In this context, the Finite Element Method (FEM) allows a wide range of investigations in terms of material properties, geometries and topology, frequency analysis, and the design phase of components. Exploring the modal behavior of soft hyperelastic structures involves studying pre-stressed conditions and analyzing how large strains or rotations affect natural frequencies and mode shapes. Existing reference solutions in this field typically focus on simple geometries or considered boundary conditions. While the literature has extensively covered various beam-like or plate-like structures, the adoption of classical FEM models, such as hexahedral solid models, is generally associated with an inadequate computational cost required by the numerical simulation. In this scenario, finite element models based on Carrera Unified Formulation (CUF) for the modal analysis of isotropic hyperelastic materials are proposed here. CUF allows, starting from the definition of the displacement field by a recursive index notation, the definition of FE governing equations in terms of invariants to the theory of structure approximation and kinematics assumption [1,2,3]. Refined fully nonlinear structural models are defined then straightforwardly [4]. The capabilities of the proposed plate CUF models are investigated through the static and modal analysis of hyperelastic thin and thick silicon plates, for which mode aberration is observed.

Hyperelastic constitutive modeling

Hyperelastic models adopted in the present work are defined under the classical strain energy function approach based on the Flory decomposition of kinematic measures. The strain energy density function, the deformation gradient $\mathbf{F}$ and the right Cauchy-Green strain tensor $\mathbf{C}$ are then written as:

$$\Psi(\mathbf{C}) = \Psi_{vol}(J) + \Psi_{iso}(\bar{\mathbf{C}}) = U(J) + \bar{\Psi}(\bar{I}_1, \bar{I}_2) \tag{1}$$

$$\mathbf{F} = \mathbf{F}_{vol}\bar{\mathbf{F}} \;\rightarrow\; \mathbf{F}_{vol} = J^{\frac{1}{3}}\mathbf{I} \quad \bar{\mathbf{F}} = J^{-\frac{1}{3}}\mathbf{F} \tag{2}$$

$$\mathbf{C} = \mathbf{C}_{vol}\bar{\mathbf{C}} \;\rightarrow\; \mathbf{C}_{vol} = J^{\frac{2}{3}}\mathbf{I} \quad \bar{\mathbf{C}} = J^{-\frac{2}{3}}\mathbf{C} \tag{3}$$

where $\bar{I}_1, \bar{I}_2$ are the invariants of the isochoric part of the right Cauchy-Green tensor $\bar{\mathbf{C}}$ and J is the volume ratio, the determinant of the deformation gradient. In the present work, the decoupled Mooney-Rivlin model for silicon rubber is taken into account

$$\Psi(\mathbf{C}) = c_{10}(\bar{I}_1 - 3) + c_{01}(\bar{I}_2 - 3) + \frac{1}{D_1}(J - 1)^2 \tag{4}$$

where $D_1 = 2/k$ is the incompressibility parameter defined from the bulk modulus k. Stress measure represented by the second Piola-Kirchoff stress tensor (PK2) is defined in its decoupled form:

$$\mathbf{S} = \frac{\partial \Psi}{\partial \mathbf{C}} = \mathbf{S}_{vol} + \mathbf{S}_{iso} = Jp\mathbf{C}^{-1} + J^{-\frac{2}{3}}\left(\mathbf{I} - \frac{1}{3}\mathbf{C}^{-1} \otimes \mathbf{C}\right):\bar{\mathbf{S}} = Jp\mathbf{C}^{-1} + J^{-\frac{2}{3}}\,\mathbf{P}:\bar{\mathbf{S}} \tag{5}$$

where $\mathbf{P}$ is the fourth-order projection tensor adopted in the Total Lagrangian Formulation finite element formulation, p is the hydrostatic pressure and $\bar{\mathbf{S}}$ is the rescaled/modified PK2 stress tensor. Due to the presence of both material and geometrical nonlinearities, an incremental formulation is here adopted. Following the procedure by Holzapfel [5], the constitutive equation Eq. (5) is rewritten in its differential form:

$$\Delta\mathbf{S} = C\frac{1}{2}\Delta\mathbf{C} = C\Delta\mathbf{E} \tag{6}$$

where C is the so-called tangent elasticity tensor, defined starting the linearization of the constitutive law. The complete derivation of the explicit expression of the tangent elasticity tensor (or material Jacobian tensor) can be found again in [5].

Higher-order structural theories
The Unified Formulation for static and modal analysis of hyperelastic soft structures has been already proposed in [6], in which hyperelastic higher-order beam finite element models are established. In the following, modal analysis of compressible hyperelastic plate structures are performed adopting higher-order plate (2D) models. In CUF, the three-dimensional displacement field is expressed as a polynomial expansion of the generalized nodal displacements, coupling approximation expansion theories along the thickness with kinematic models along the plate mid-surface:

$$\mathbf{u}(x, y, z) = F_\tau(z)\mathbf{u}_\tau(x, y) = F_\tau(z)N_i(x, y)\mathbf{u}_{\tau i} \quad \tau = 1, \dots, M, \quad i = 1, \dots, N_n \tag{8}$$

where M is the related to the order of the structural theory adopted, $F_\tau(z)$ is the set of expansion functions, representing the theory of structure approximation, $N_i(x, y)$ is the set of 2D shape functions of the discrete N_n finite nodes along the mid-surface, and finally $\mathbf{u}_{\tau i}$ is the vector of generalized displacement component.. In the present work, Lagrange Expansion (LE) class are considered, starting from the set of Lagrange's polynomials [2].

Governing equations
The equilibrium equations for the static and undamped vibration problem are carried out by means of the Principle of Virtual Displacements (PVD), written as:

$$\delta\mathcal{L}_{ine} + \delta\mathcal{L}_{int} = \delta\mathcal{L}_{ext} \tag{9}$$

where $\mathcal{L}_{ine}$ is the work done by inertial forces, $\mathcal{L}_{int}$ and $\mathcal{L}_{ext}$ represents work done by internal and external forces, and δ denotes the virtual variation. Adopting the same indices notation introduced for the displacement field also for the Green-Lagrange strain tensor and the virtual quantities, one can derive the FN (Fundamental Nuclei) of the internal and external forces vector and mass matrix, obtaining the matrix form of the governing equation.

$$\delta u_{sj}: \; M^{\tau sij}\ddot{u}_{\tau i} + F_{int}^{sj} = F_{ext}^{sj} \;\rightarrow\; \mathbf{M}\ddot{u} + F_{int} = F_{ext} \tag{9}$$

The nonlinear governing equation is then linearized to implement an incremental numerical solver based on path-following constraints. Through a Taylor expansion truncated at the first order, one can derive the incremental equation:

$$M\Delta\ddot{u} + K_T\Delta u = -\phi_{res}(u_0, \dot{u}_0, p_0) + I\Delta\lambda p_{ref} \tag{9}$$

Finally, imposing a harmonic solution of the type $\Delta u = \Phi e^{i\omega t}$, the classical linear eigenvalue problem is then obtained, that gives the natural frequencies and the normal modes of vibration around the computed non-trivial equilibrium state:

$$(K_T - \omega^2 M)\Phi = 0 \tag{9}$$

More detail about the derivation of the undamped vibration problem around non-trivial equilibrium conditions of hyperelastic soft structures can be found in [7], where the FN of the tangent stiffness matrix and the complete linearization procedure is detailed.

Numerical results

The investigated case study involves a clamped square plate of compressible hyperelastic material. Small amplitude vibrations around non-trivial equilibrium conditions are explored considering a thick square plate with a lateral side of a=b=1 and h=0.1 m. In the further investigations, the material density of the hyperelastic beam is set to a typical value for silicone rubber $\rho = 1200$ kg/m^3. The fitted material parameters of the Mooney-Rivlin model for silicon rubber are fixed to $c_{10} = 0.14$ MPa and $c_{01} = 0.023$ MPa [7], and Poisson's ratio $\nu = 0.4$. The mathematical models utilized different LE models along the plate thickness, namely the LE2 parabolic and LE3 cubic expansion models, and various finite element discretizations around the plate mid-surface to assess the accuracy and efficiency of the proposed model. The relative percentage difference and the computational costs in terms of DOF (degrees of freedom) will be further presented. The numerical results obtained via higher-order 2D CUF plate elements are compared with the 3D solution carried out through ABAQUS commercial software. First, a free vibration analysis around the undeformed equilibrium condition is performed to analyze the natural frequencies and modal shapes of the thick plate. Table 1 shows the first five natural frequencies, comparing the results with the solution obtained with the 3D model in ABAQUS. The most accurate model involves 20x20 Q9 elements along the mid-surface, which will be further adopted as discretization. Furthermore, the nonlinear static analysis is performed, considering a uniform pressure at the top surface to establish nontrivial equilibrium conditions. Figure (1) illustrates the equilibrium curve obtained using higher-order 2D CUF elements and the 3D ABAQUS solution. Around the marked nontrivial equilibrium conditions, the undamped vibration problem is solved, computing the natural frequencies for increasing applied pressure values. Figure (2) presents the first eight natural frequencies and their variations with increasing applied pressure, with results compared to the proposed 3D solution. Accurate results are consistently achieved, and mode aberration is observed.

Conclusions

This manuscript discusses the undamped vibration problem around non-trivial equilibrium conditions of hyperelastic plate structures. First, soft hyperelastic plates have been modeled using higher-order CUF-based finite elements and have proven to guarantee accurate results regarding displacements and modal behavior of thin and thick structures. Mode aberration is observed, such as crossing, as observed in the proposed study, for specific critical values of the applied load. Future works will investigate hyperelastic multilayered soft plates and shells, the modal analysis of bio-inspired structures, and the effect of fiber reinforcement.

Table 1: Cantilever compressible silicon thick plate: free vibration problem around the undeformed condition, convergence analysis on the first five natural frequencies [Hz]. Comparison between 2D CUF results and 3D ABAQUS results.

Mesh	Mode 1	Mode2	Mode 3	Mode 4	Mode 5	DOFs
10x10 Q9	$0.4421^{0.417\%}$	$1.0421^{0.809\%}$	$2.6030^{0.876\%}$	$2.7948^{0.156\%}$	$3.2900^{0.629\%}$	3969
12x12 Q9	$0.4418^{0.003\%}$	$1.0411^{0.708\%}$	$2.6002^{0.768\%}$	$2.7937^{0.115\%}$	$3.2880^{0.568\%}$	5625
15x15 Q9	$0.4415^{0.271\%}$	$1.0403^{0.626\%}$	$2.5979^{0.682\%}$	$2.7927^{0.079\%}$	$3.2866^{0.524\%}$	8649
20x20 Q9	$0.4412^{0.211\%}$	$1.0396^{0.565\%}$	$2.5962^{0..616\%}$	$2.7919^{0.051\%}$	$3.2856^{0.495\%}$	15129
ABQ 3D 12500 C3D20R	0.4403	1.0338	2.5804	2.7905	3.2695	177633

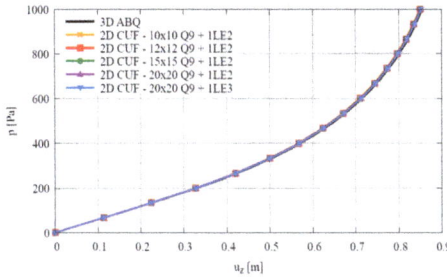

Figure 1: Cantilever compressible silicon thick plate: equilibrium paths

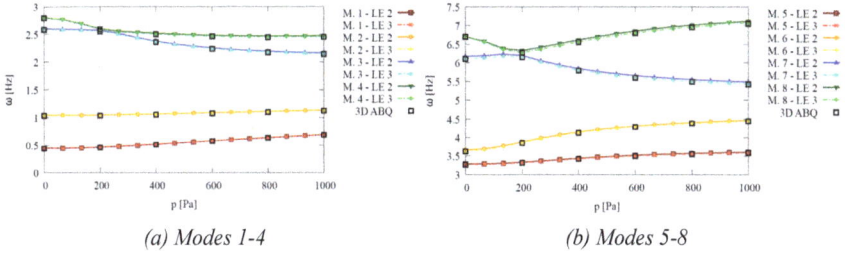

(a) Modes 1-4	*(b) Modes 5-8*

Figure 2: Cantilever compressible silicon plate, case L/h = 10: variation of the first eight natural frequencies for increasing value of the applied pressure.

References

[1] E. Carrera, A. Pagani, R. Azzara, R. Augello. Vibration of metallic and composite shells in geometrical nonlinear equilibrium states. Thin-Walled Structures, 157:107131, dec 2020. http://doi.org/10.1016/j.tws.2020.107131

[2] E. Carrera, M. Cinefra, E. Zappino, M. Petrolo. Finite Element Analysis of Structures Through Unified Formulation. Wiley, Chichester, West Sussex, UK, 2014, jul 2014. ISBN: 9781118536643.

Aerospace Science and Engineering - III Aerospace PhD-Days   Materials Research Forum LLC
Materials Research Proceedings 42 (2024) 26-30   https://doi.org/10.21741/9781644903193-7

[3] E. Carrera, M. Cinefra , M. Petrolo, E. Zappino. Comparisons between 1D (beam) and 2S (plate/shell) finite elements to analyze thin walled structures. Aerotecnica Missili & Spazio, 93(1–2):3–16, January 2014. http://doi.org/10.1007/BF03404671

[4] A. Pagani, E. Carrera. Unified formulation of geometrically nonlinear refined beam theories. Mechanics of Advanced Materials and Structures, 25(1):15–31, sep 2016. http://doi.org/10.1080/15376494.2016.1232458

[5] G.A. Holzapfel. Nonlinear Solid Mechanics. John Wiley & Sons, Chichester, West Sussex, UK, 2000.

[6] A. Pagani, P. Chiaia, E. Carrera. Vibration of solid and thin-walled slender structures made of soft materials by high-order beam finite elements. International Journal of Non-Linear Mechanics, 160:104634, April 2024. http://doi.org/10.1016/j.ijnonlinmec.2023.104634

[7] L. Meunier, G. Chagnon, D. Favier, L. Orgeas, P. Vacher. Mechanical experimental characterisation and numerical modelling of an unfilled silicone rubber. Polymer Testing, 27(6):765–777, September 2008. http://doi.org/10.1016/j.polymertesting.2008.05.011

Aerospace Science and Engineering - IV Aerospace PhD-Days   Materials Research Forum LLC
Materials Research Proceedings 42 (2024) 31-37   https://doi.org/10.21741/9781644903193-8

Novel approach to damage detection in laminated structures: modal damping as a damage indicator

Shabnam Kiasat[1,a*]

[1] Mul² Lab, Department of Mechanical and Aerospace Engineering, Politecnico di Torino, 10129, Torino, Italy

[a]shabnam.kiasat@polito.it

Keywords: Modal Damping, Delamination, SHM, Damage Detection, Viscoelasticity, Contact Damping

Abstract. This paper provides a general overview of the author's Ph.D. research. It emphasizes the criticality of damage detection in composite materials, focusing on interlaminar delamination. It introduces vibration-based non-destructive testing (NDT) techniques and modal damping analysis as a novel damage indicator and highlights their real-time capabilities and sensitivity to subtle defects. Numerical modeling benefits and challenges in understanding modal damping behavior are discussed. The review covers research on modal damping modeling, its application in detecting interlaminar delamination, and composite defect simulation. The findings of the presented study provide insights into delamination behavior and its impact on structural integrity. Overall, this work highlights the effectiveness of vibration-based NDT and numerical modeling for enhanced structural health monitoring and material safety.

Introduction: Challenges in Structural Integrity and the Imperative for Damage Detection

Composite materials, extensively utilized across industries including aerospace [1], automotive [2], and railways [3], exhibit exceptional characteristics such as a favorable balance between strength and weight, resistance to corrosion, and adaptability in design.

Defects in composites: Laminated structures are vulnerable to a range of damage modes influenced by factors such as initial geometric imperfections [4], mechanical properties, boundary conditions, and applied loads [5]. These failure mechanisms are typically categorized based on the location of the defect, distinguishing between intralaminar [6] and interlaminar [7] damages. Intralaminar damage, such as matrix cracking, fiber fracture, fiber/matrix debonding, and fiber pull-out [8], primarily manifests at the free edges of the structure due to stress concentration in these areas. Conversely, interlaminar defects, such as delamination and interlaminar shear cracking, occur within the bulk of the laminate and are often associated with geometric irregularities introduced during manufacturing processes or because of impact loading. One prevalent interlaminar defect is interlaminar delamination (hereinafter, delamination), characterized by a separation between adjacent layers of the laminate [9]. Delamination can occur over a large area or involve multiple regions, posing a significant risk of catastrophic structural failure if left undetected or untreated. Understanding the necessity of detecting subcritical damage initiation in laminates is crucial for ensuring safety and maintenance in composite structures.

The benefits of NDT techniques for damage detection: Non-destructive testing (NDT) techniques play a crucial role in upholding structural integrity and safety in diverse industries, with a particular emphasis on composites and laminated structures. These techniques are indispensable for ensuring the integrity and safety of structures in a range of industries [10]. These techniques offer numerous benefits that are essential for ensuring structural integrity and safety across various industries. Firstly, NDT methods facilitate early defect detection by analyzing material characteristics, internal structures, and surface conditions without causing any impact on the object itself. This proactive approach enables timely maintenance, minimizing downtime, and mitigating

Aerospace Science and Engineering - III Aerospace PhD-Days   Materials Research Forum LLC
Materials Research Proceedings 42 (2024) 31-37   https://doi.org/10.21741/9781644903193-8

the risk of catastrophic failures, which is especially crucial in the case of composite materials prone to delamination or fiber breakage. Secondly, these techniques lead to substantial cost savings by identifying damages at an early stage, thus averting expensive replacements or extensive repairs in composite structures. Thirdly, NDT assures safety by ensuring that composite components meet quality standards and specifications, thereby contributing to safer operations, particularly in critical sectors like aerospace and marine engineering where laminated structures are prevalent [11]. Additionally, NDT offers non-intrusive inspection, preserving the integrity of composite materials during examination, which is crucial for in-service assessments. Lastly, NDT techniques possess wide applicability in the composite industry, capable of detecting both internal and external irregularities, determining material composition, and making precise measurements across various composite materials and structures. In summary, NDT methods serve as essential tools for maintaining safety, prolonging the lifespan of composite structures, and ensuring reliable performance. By integrating different NDT techniques, engineers can conduct comprehensive assessments, particularly in the realm of composite materials and laminated structures [12].

Advanced Composite Defect Detection: Vibration-Based NDT and Modal Damping Analysis
Vibration-based NDT techniques: Unlike most current NDT techniques, which often require out-of-service, offline measurements, oscillation-based techniques offer the advantage of real-time, online applications. Methods like Ultrasound or Acoustic Emission necessitate the equipment to be taken offline for assessment, interrupting operations and potentially delaying defect detection. Conversely, oscillation-based techniques, including Lamb waves and vibration-based Structural Health Monitoring (SHM), enable the detection of damage in real time, facilitating continuous monitoring without disrupting operations. Vibration-based SHM techniques, in particular, present logistical simplicity and cost-effectiveness, making them attractive for widespread implementation. However, it's worth noting that while vibration-based techniques are generally sensitive to large damages, they may exhibit insensitivity to smaller defects [13]. Yet, modal damping, a characteristic of the structure inherent in vibration-based methods, offers sensitivity to subtle damages such as small delamination, providing a nuanced understanding of structural health [14]. Therefore, despite some limitations, vibration-based NDT techniques, especially those incorporating global modal parameters, offer valuable insights into structural integrity in real-time and at a relatively low cost. Before delving into the numerical modeling challenges, it's crucial to understand the behavior of modal damping as accurate predictions are essential for effective damage detection strategies.

Modal damping as a Damage Index (DI): In frequency domain data analysis, identifying structural deficiencies often involves observing alterations in modal parameters, including natural frequency [15, 16], mode shape curvature [17, 18], and modal damping [19, 20]. However, previous research has noted that despite its global nature, natural frequency is generally insensitive to small damages [14]. Similarly, the localized nature of mode shape curvature makes it less suitable for detecting defects occurring in arbitrary locations within structures. The prior experimental research serves as the background for this study, as it demonstrated significant variations in modal damping based on the extent of delamination within a composite structure [14]. Therefore, this study aims to explore modal damping as a reliable damage indicator (DI) in delaminated structures.

Numerical modeling of modal damping: Given the potential of modal damping as a DI in delaminated structures, Utilizing numerical modeling techniques is crucial for gaining a deeper understanding of modal damping behavior in delaminated structures. The utilization of numerical models not only offers numerous advantages but also presents certain challenges that need to be addressed for accurate and reliable predictions. Numerical modeling of modal damping provides several advantages, including the ability to simulate a wide range of loading conditions and damage scenarios [21]. Through numerical simulations, researchers can systematically investigate the influence of various factors, such as material properties, geometric configurations, and environmental conditions, on modal damping [22]. Additionally, numerical models enable the efficient exploration of complex structural behaviors, facilitating the identification of optimal monitoring strategies and damage detection algorithms. Moreover, numerical simulations offer cost-effective alternatives to experimental testing, allowing for extensive parametric studies and sensitivity.

Exploring Modal Damping Modeling and Composite Defect Simulation: Insights from Current Methods

An Overview of Numerical Methods for Modal Damping Modeling: Numerical modeling plays a significant role in studying modal damping, a crucial parameter in structural dynamics that governs the rate at which mechanical energy dissipates in vibrating structures. It offers a powerful toolset to accurately analyze and predict damping characteristics. Modal damping, defined as the ratio of the energy dissipated per cycle to the total energy of the vibrating system, characterizes the structure's ability to absorb and dissipate energy during vibration. Various factors, including material properties, structural geometry, boundary conditions, and environmental effects, influence modal damping. In structural dynamics analyses, modal damping is typically represented using damping ratios associated with each mode of vibration [23]. Regarding numerical techniques for analyzing modal damping behavior in structural dynamics, the Finite Element Method (FEM) is widely utilized. FEM discretizes the structure into finite elements, enabling accurate representation of complex geometries and material properties [24]. While traditional eigenvalue analysis typically neglects damping effects, specialized techniques incorporating damping models such as Rayleigh damping [25], Prony series [26], or material damping [27] formulations can be employed within FEM analyses to account for modal damping behavior. Modal analysis techniques are then employed to extract modal parameters, including natural frequencies, mode shapes, and modal damping ratios.

Although numerical algorithms such as Lanczos iteration [28] or subspace iteration methods [29] are commonly used to solve the undamped eigenvalue problem, they may not fully capture the effects of damping. In contrast, time-domain simulations [30] offer insights into dynamic responses and modal damping behavior under transient loading conditions, providing a more comprehensive understanding of damping behavior. While time-domain simulations are computationally intensive, they offer the advantage of capturing nonlinear and transient effects that may affect modal damping. Reduced Order Modeling (ROM) techniques such as Craig-Bampton or Guyan reduction are employed to mitigate computational costs while maintaining accuracy, generating simplified models that retain only dominant modes of vibration, significantly reducing computational burdens [31]. Despite the advantages of numerical modeling, challenges exist in accurately predicting modal damping behavior. These challenges include numerical damping artifacts, model validation against experimental data, and computational efficiency [13]. Careful consideration of numerical methods and parameters is essential to ensure reliable predictions of modal damping. In experimental works, it is challenging to investigate an exclusive damping mechanism, as multiple damping mechanisms are often present [14]. Therefore, experimental validation of damping models may involve complexities in isolating and characterizing individual damping contributions, further emphasizing the need for comprehensive numerical modeling approaches. Numerical methods have been extensively applied to model modal damping in various engineering applications. Case studies demonstrate the effectiveness of numerical techniques in predicting modal damping behavior and guiding engineering decisions, including aerospace structures, automotive components, and civil engineering infrastructure subjected to dynamic loading conditions [32, 33]. In conclusion, numerical methods play a crucial role in modeling modal damping behavior, offering powerful tools for analyzing and predicting damping characteristics in vibrating structures.

Aerospace Science and Engineering - III Aerospace PhD-Days   Materials Research Forum LLC
Materials Research Proceedings 42 (2024) 31-37   https://doi.org/10.21741/9781644903193-8

Multi-Scale Modeling of Defects in Laminate Structures: Multi-scale modeling of interlaminar and intralaminar defects in composite laminates is indispensable for understanding and predicting structural performance [34]. These laminates, comprising stacked layers of fibers and matrix, exhibit complex behavior influenced by defects across various scales. At the micro-scale, intralaminar defects such as matrix cracking, fiber breakage, and fiber/matrix debonding occur within individual layers [35]. Microscopic finite element models are employed to capture these defects, considering interactions between fibers and matrix, thus aiding in predicting local stress concentrations and failure initiation sites [36]. Additionally, interlaminar defects, such as micro-voids embedded within fiber bundles, significantly impact mechanical properties but are challenging to detect. Peridynamic models at this scale analyze void-induced stiffness reduction and crack initiation near voids [37, 38]. Transitioning to the mesoscale, models focus on individual layers or ply interfaces, addressing transverse matrix cracking and delamination between adjacent layers [39, 40]. These models consider the interaction between neighboring plies and the effect of defects on stiffness and strength. Furthermore, macro-scale finite element models simulate entire composite structures, incorporating intralaminar defects such as matrix cracks and fiber breakage, thus predicting global stiffness reduction and load-carrying capacity. Meanwhile, delamination at the macro-scale affects overall laminate strength and stability, with cohesive zone models [41] employed to simulate crack propagation along layer interfaces, considering energy release rates and fracture toughness. Comparing these modeling approaches, micro-scale models provide detailed insights into local defect behavior but are computationally expensive. Meso-scale models offer a balance between detail and efficiency, capturing both intralaminar and interlaminar defects, although they may still require significant computational resources. Macro-scale models offer a global view of structural behavior but may oversimplify defect interactions. Cohesive zone models excel at capturing crack propagation but require accurate input parameters. In conclusion, each modeling scale offers unique advantages and challenges. Integrating multi-scale modeling approaches can provide a comprehensive understanding of defect behavior in composite laminates, aiding in material design and structural optimization. However, careful consideration of computational resources, model complexity, and validation against experimental data is crucial to ensure accurate and reliable predictions.

Conclusion

This work provides an overview of the author's Ph.D. research which emphasizes the importance of detecting interlaminar delamination defects in laminate structures to ensure their integrity. By employing vibration-based non-destructive testing (NDT) techniques with modal damping analysis, we accurately identified these critical defects. Supported by numerical modeling, particularly the finite element method (FEM), we analyzed the structure's behavior, considering viscoelastic (VE) and frictional damping mechanisms. We performed dynamic explicit analysis and utilized fast Fourier transform (FFT) for time-to-frequency conversion. Parametric studies rigorously compared defected and pristine structures, assessing changes in resonance frequencies and modal damping. These findings contribute significantly to advancing structural health monitoring (SHM) practices, providing invaluable insights for assessing and maintaining composite structures in service.

References

[1] A. Vedernikov, A. Safonov, F. Tucci, P. Carlone, I. Akhatov, Pultruded materials and structures: A review, Journal of Composite Materials. 54 (2000) 4081–4117. https://doi.org/10.1177/0021998320922894

[2] G. I. Zagainov, G. E. Lozino-Lozinsky, Composite Materials in Aerospace Design, Springer Science & Business Media, 1996. https://doi.org/10.1007/978-94-011-0575-0

Aerospace Science and Engineering - III Aerospace PhD-Days   Materials Research Forum LLC
Materials Research Proceedings 42 (2024) 31-37   https://doi.org/10.21741/9781644903193-8

[3] D. K. Rajak, D. D. Pagar, A. Behera, P. L. Menezes, Role of composite materials in automotive sector: potential applications, Advances in Engine Tribology. (2000) 193-217. https://doi.org/10.1007/978-981-16-8337-4_10

[4] L. Lui, B. Z. Zhang, D. F. Wang, & Z. J. Wu, Effects of cure cycles on void content and mechanical properties of composite laminates, Composite Structures. 73 (2006) 303-309. https://doi.org/10.1016/j.compstruct.2005.02.001

[5] E. Benvenuti, O. Vitarelli, A. Tralli, Delamination of FRP-reinforced concrete by means of an extended finite element formulation, Composites Part B: Engineering. 43 (2012) 3258–3269. https://doi.org/10.1016/j.compositesb.2012.02.035

[6] S. Hühne, J. Reinoso, E. Jansen, R. Rolfes, A two-way loose coupling procedure for investigating the buckling and damage behavior of stiffened composite panels, Composite Structures. 136 (2016) 513–525. https://doi.org/10.1016/j.compstruct.2015.09.056

[7] Allix, O., & Ladevèze, P., Interlaminar interface modeling for the prediction of delamination, Composite Structures. 22 (1992) 235–242. https://doi.org/10.1016/0263-8223(92)90060-P

[8] M. Saeedifar, H. Hosseini Toudeshky, The effect of interlaminar and intralaminar damage mechanisms on the quasi-static indentation strength of composite laminates, Applied Composite Materials. 30 (2023) 871–886. https://doi.org/10.1007/s10443-023-10123-x

[9] S. A. Ramu, V. T. Johnson, Damage assessment of composite structures—a fuzzy logic integrated neural network approach, Computers & Structures. 57 (1995) 491–502. https://doi.org/10.1016/0045-7949(94)00624-C

[10] A. M. Amaro, P. N. B. Reis, M. F. S. F. De Moura, J. B. Santos, Damage detection on laminated composite materials using several NDT techniques, Insight-Non-Destructive Testing and Condition Monitoring. 54 (2012) 14-20. https://doi.org/10.1784/insi.2012.54.1.14

[11] P.E. Mix, Introduction to Nondestructive Testing: A Training Guide, 2nd ed., John Wiley & Sons, 2005. https://doi.org/10.1002/0471719145

[12] P. Duchene, S. Chaki, A. Ayadi, P. Krawczak, A review of non-destructive techniques used for mechanical damage assessment in polymer composites, Journal of Materials Science. 53 (2018) 7915-7938. https://doi.org/10.1007/s10853-018-2045-6

[13] A. S. Nobari, M. H. F. Aliabadi, Vibration-based Techniques for Damage Detection and Localization in Engineering Structures, World Scientific, 2018. https://doi.org/10.1142/9781786344977_0001

[14] M. Khazaee, A.S. Nobari, M.H.F. Aliabadi, Experimental Investigation of Delamination Effects on Modal Damping of a CFRP Laminate, Using a Statistical Rationalization Approach, in: Vibration-Based Techniques for Damage Detection and Localization in Engineering Structures, World Scientific, 2018, pp. 75-103. https://doi.org/10.1142/9781786344977_0003

[15] A. Paolozzi, I. Peroni, Identification of models of a composite plate for the purpose of damage detection, AEROTEC. MISSILI SPAZIO. 6 7(1988) 119-129.

[16] A. Paolozzi, I. Peroni, Detection of debonding damage in a composite plate through natural frequency variations, Journal of Reinforced Plastics and Composites. 9 (1990) 369-389. https://doi.org/10.1177/073168448400090040

[17] A. K. Pandey, M. Biswas, M. M. Samman, Damage detection from changes in curvature mode shapes, Journal of Sound and Vibration. 145(1991) 321-332. https://doi.org/10.1016/0022-460X(91)90595-B

[18] P. Qiao, K. Lu, W. Lestari, J. Wang, Curvature mode shape-based damage detection in composite laminated plates, Composite Structures. 80(2007) 409-428. https://doi.org/10.1016/j.compstruct.2006.05.026

[19] B. Nanda, D. Maity, D. K. Maiti, Vibration-based structural damage detection technique using particle swarm optimization with incremental swarm size, International Journal Aeronautical and Space Sciences. 13(2012) 323-331. https://doi.org/10.5139/IJASS.2012.13.3.323

[20] T. R. Jebieshia, D. Maiti, Vibration characteristics and damage detection of composite structures with anisotropic damage using unified particle swarm optimization technique. In: Proceedings of the American Society for Composites, 13th Technical Conference, East Lansing, MI, 2015, pp. 28–30. https://doi.org/10.1016/j.compscitech.2015.04.014

[21] D. A. Pereira, T. A. M. Guimarães, H. B. Resende, D. A. Rade, Numerical and experimental analyses of modal frequency and damping in tow-steered CFRP laminates, Composite Structures. 244 (2020) 112-190. https://doi.org/10.1016/j.compstruct.2020.112190

[22] S. E. Hamdi, Z. M. Sbartaï, S. M. Elachachi, Performance assessment of modal parameters identification methods for timber structures evaluation: Numerical modeling and case study, Wood Science and Technology. 55 (2021) 1593-1618. https://doi.org/10.1007/s00226-021-01335-0

[23] S. Adhikari, Structural Dynamic Analysis with Generalized Damping Models: Analysis, John Wiley & Sons, 2013. https://doi.org/ 10.1142/9781786344977_0001

[24] E. Carrera, M. Cinefra, M. Petrolo, E. Zappino, Finite Element Analysis of Structures Through Unified Formulation, John Wiley & Sons, 2014. https://doi.org/10.1002/9781118572023

[25] R. E. Spears, S. R. Jensen, Approach for selection of Rayleigh damping parameters used for time history analysis, The Electronic Journal of Geotechnical Engineering. 6 (2012) 061801. https://doi.org/10.1115/1.4006855

[26] Sh. Kiasat, M. Filippi, A. S. Nobari, E. Carrera, Layer-wise dynamic analysis of a beam with global and local viscoelastic contributions using an FE/Laplace transform approach, Acta Mechanica. 233 (2022) 4747-4761. https://doi.org/10.1007/s00707-022-03349-6

[27] C. W. Bert, Material damping: An introductory review of mathematical models, measures, and experimental techniques, Journal of Sound and Vibration. 29 (1973) 129-153. https://doi.org/10.1016/S0022-460X(73)80131-2

[28] J. K. Cullum, R. A. Willoughby, Lanczos Algorithms for Large Symmetric Eigenvalue Computations, SIAM, 2002. https://doi.org/10.1007/978-1-4684-9192-0

[29] K. J. Bathe, The subspace iteration method–Revisited, Computers & Structures 126 (2013) 177-183. https://doi.org/10.1016/j.compstruc.2012.06.002

[30] X. Y. Li, S. S. Law, Identification of structural damping in time domain, Journal of Sound and Vibration. 328 (2009) 71-84. https://doi.org/10.1016/j.jsv.2009.07.033

[31] M. Kurt, Identification, Reduced Order Modeling and Model Updating of Nonlinear Mechanical Systems, University of Illinois at Urbana-Champaign, 2014. https://doi.org/10.1142/9781118572023

[32] G. Kerschen, M. Maxime, J. C. Golinval, C. Stephan, Nonlinear modal analysis of a full-scale aircraft, Journal of Aircraft, 5(2013) 1409–1419. https://doi.org/10.2514/1.C031918

Aerospace Science and Engineering - III Aerospace PhD-Days   Materials Research Forum LLC
Materials Research Proceedings 42 (2024) 31-37   https://doi.org/10.21741/9781644903193-8

[33] G. A. Papagiannopoulos, D. E. Beskos, On a modal damping identification model of building structures, Archive of Applied Mechanics. 76 (2006) 443-463. https://doi.org/10.1007/s00419-006-0046-4

[34] K.-N. Antin, A. Laukkanen, T. Andersson, D. Smyl, P. Vilaça, A multiscale modelling approach for estimating the effect of defects in unidirectional carbon fiber reinforced polymer composites, Materials. 12 (2019) 1885. https://doi.org/10.3390/ma12121885

[35] T. A. Sebaey, G. Catalanotti, N. P. O'Dowd, A microscale integrated approach to measure and model fiber misalignment in fiber-reinforced composites, Composites Science and Technology. 183 (2019) 107793. https://doi.org/10.1016/j.compscitech.2019.107793

[36] Y. Fu, X. Yao, X. Gao, Micro-mesoscopic prediction of void defect in 3D braided composites, Composites Part A: Applied Science and Manufacturing. 147(2021) 106450. https://doi.org/10.1016/j.compositesa.2021.106450

[37] A. Yaghoobi, M. G. Chorzepa, S. S. Kim, S. A. Durham, Mesoscale fracture analysis of multiphase cementitious composites using peridynamics, Materials. 10 (2017) 162. https://doi.org/10.3390/ma10020162

[38] M. Zaccariotto, G. Sarego, D. Dipasquale, et. al. Discontinuous mechanical problems studied with an aerodynamics-based approach, Aerotecnica Missili & Spazio. 96 (2017) 44-55. https://doi.org/10.1007/BF03404736

[39] Y. Yu, J. Ye, Y. Wang, B. Zhang, G. Qi, A mesoscale ultrasonic attenuation finite element model of composites with random-distributed voids, Composites Science and Technology. 89 (2013) 44-51. https://doi.org/10.1016/j.compscitech.2013.09.006

[40] S. Kiasat, A. S. Nobari, M. Filippi, E. Carrera, Characterization of viscoelastic damping mechanism in a delaminated structure, Mechanical Systems and Signal Processing. (Manuscript submitted for publication).

[41] S. A. Ponnusami, S. Turteltaub, S. van der Zwaag, Cohesive-zone modelling of crack nucleation and propagation in particulate composites, Engineering Fracture Mechanics 149 (2015) 170-190. https://doi.org/10.1016/j.engfracmech.2015.09.050

Aerospace Science and Engineering - IV Aerospace PhD-Days

Materials Research Proceedings 42 (2024) 38-42

Materials Research Forum LLC

https://doi.org/10.21741/9781644903193-9

Continuum sensitivity analysis and improved Nelson's method for beam shape eigensensitivities

Giuseppe Maurizio Gagliardi[1,a] *, Mandar D. Kulkarni[2,b] and Francesco Marulo[1,c]

[1]Department of Industrial Engineering, University of Naples Federico II, Via Claudio, 21, Naples, 80125, Italy

[2]Department of Aerospace Engineering, Embry-Riddle Aeronautical University, 1 Aerospace Boulevard, Daytona Beach, 32114, Florida, USA

[a]giuseppemaurizio.gagliardi@unina.it, [b]kulkarnm@erau.edu, [c]francesco.marulo@unina.it

Keywords: Continuum Sensitivity Analysis, Nelson's Sensitivity Method, Gradient-Based Optimization Techniques, Shape Eigensensitivity

Abstract. Gradient-based optimization techniques require accurate and efficient sensitivity or design derivative analysis. In general, numerical sensitivity methods such as finite differences are easy to implement but imprecise and computationally inefficient. In contrast, analytical sensitivity methods are highly accurate and efficient. Although these methods have been widely evaluated for static problems or dynamic analysis in the time domain, no analytical sensitivity methods have been developed for eigenvalue problems. In this paper, two different analytical methods for shape eigensensitivity analysis have been evaluated: the Continuum Sensitivity Analysis (CSA) and an enhanced version of Nelson's method. They are both analytical techniques but differ in how the analytical differentiation is performed: before and after the discretization, respectively. CSA has been applied to eigenvalue problems for the first time, while Nelson's method has been improved and adapted to shape optimizations. Both methods have been applied to different cases involving shape optimization of beams. Both vibration and buckling problems were analysed considering the eigenvalue as a design variable. Both methods have been successfully applied, and Nelson's method proved to be more convenient for this kind of problem.

Introduction

Accurate sensitivity analysis is essential to guarantee the convergence of gradient-based optimization techniques. The sensitivity methods can be divided into numerical methods (finite difference, complex step), analytical methods (discrete analytical, continuum), hybrid methods (semi-analytical), or automatic differentiation methods. Analytical methods are preferred over numerical ones because of their higher accuracy and computational efficiency. They do not require convergence studies to find an adequate step size for calculating the numerical derivative, as for the finite difference method and the semi-analytical one [1,2,3]. Furthermore, they do not need the source code of the analysis or to handle complex operations, as needed for automatic differentiation [4,5] or complex step one [6], respectively. Analytical methods offer an accurate and efficient alternative to compute derivatives for structures [7,8,9], fluids [10,11,12], and fluid-structure-interaction problems [13,14] with respect to shape design parameters. Continuum Sensitivity Analysis (CSA) has been developed to compute gradients to be used in shape optimizations for static structural problems or dynamic problems in the time domain. In this work, it has been extended for the first time to eigenvalue problems. On the other hand, Nelson's method has been widely applied for calculating derivatives with respect to design parameters. In this paper, it has been extended for the first time to shape optimizations. Both methodologies are successfully developed and validated in this paper. However, CSA exhibited some limitations in terms of range of applicability and accuracy. Instead, Nelson's method exhibits very good accuracy and

Aerospace Science and Engineering - III Aerospace PhD-Days   Materials Research Forum LLC
Materials Research Proceedings 42 (2024) 38-42   https://doi.org/10.21741/9781644903193-9

computational efficiency even with coarse meshes. In particular, a nonintrusive and element-agnostic approach was pursued for the method to be suitable for standard commercial software in a black box configuration. Because of that, the approach can be adopted for practical applications concerning structural shape optimization.

Differentiation of the analytical eigenvalue problem

Considering the structural free vibration or buckling problem, the formal operator equation of the eigenvalue problem can be described by the following general equation:

$$Ay = \zeta By, y \neq 0, \tag{2}$$

where y is used to indicate the eigenfunction, and ζ is the associated eigenvalue. The following normalization condition has been employed to scale the eigenfunctions:

$$(By, y) = 1, \tag{3}$$

where $(\,,\,)$ indicates the L_2-scalar product. The principle of virtual work can be applied to get the variational formulation of the eigenvalue problem. The L_2-scalar product on both sides of Eq. (3) with a smooth function $\overline{y}$ satisfying the same boundary conditions as y may be used to obtain the variational equation of the eigenvalue problem as:

$$a(y, \overline{y}) \equiv (Ay, \overline{y}) = \zeta(By, \overline{y}) \equiv \zeta d(y, \overline{y}). \tag{4}$$

A vibrating structure eigenvalue ζ_τ on a deformed domain Ω_τ is determined by a variational equation of the form:

$$
\begin{aligned}
a_{\Omega_\tau}(y_\tau, \overline{y}_\tau) &\equiv \iint_{\Omega_\tau} c(y_\tau, \overline{y}_\tau) d\Omega_\tau \\
&= \zeta_\tau \iint_{\Omega_\tau} e(y_\tau, \overline{y}_\tau) d\Omega_\tau \equiv \zeta_\tau d_{\Omega_\tau}(y_\tau, \overline{y}_\tau), \forall \overline{y}_\tau \in Z_\tau,
\end{aligned} \tag{5}
$$

where Z_τ is the space of kinematically admissible displacements, and $c(\,,\,)$ and $e(\,,\,)$ are symmetric bilinear mappings. Since Eq. 5 is homogeneous in eigenfunction y_τ, a normalizing condition must be used to define a unique solution. The following is the one used in this discussion:

$$d_{\Omega_\tau}(y_\tau, \overline{y}_\tau) = 1. \tag{6}$$

The rigorous derivation of such an eigenvalue problem with respect to the shape, due to Choi and Kim [9], brings to the following general formula:

$$
\begin{aligned}
\zeta' &= 2 \iint_\Omega [-c(y, \nabla yV) + \zeta e(y, \nabla y)] d\Omega + \int_\Gamma [c(y, y) - \zeta e(y, y)] V_n d\Gamma \\
&= 2 \iint_\Omega [-c(y, \nabla yV) + \zeta e(y, \nabla yV)] d\Omega + \iint_\Omega \text{div}([c(y, y) - \zeta e(y, y)]V) \, d\Omega.
\end{aligned} \tag{7}
$$

This formula can be particularized and simplified based on the type of modes and design velocities.

Differentiation of the numerical eigenvalue problem

Nelson's method is a discrete analytical sensitivity method and requires the governing equations first, to be discretized and second, to be differentiated. This kind of method involves the derivatives of the Finite Element (FE) matrices. Given the symmetric real matrices $[K]$, $[M]$, $[K']$ and $[M']$ $\in \mathbb{R}^{n \times n}$, where $[K'] \equiv \frac{\partial [K]}{\partial p}$ and $[M'] \equiv \frac{\partial [M]}{\partial p}$ with p shape parameter, let $\lambda \in \mathbb{R}$ and $\{x\} \in \mathbb{R}^n$ solve the following generalized eigenvalue problem:

$$[K]\{x\} = \lambda [M]\{x\}. \tag{8}$$

Aerospace Science and Engineering - III Aerospace PhD-Days   Materials Research Forum LLC
Materials Research Proceedings 42 (2024) 38-42   https://doi.org/10.21741/9781644903193-9

The numerical eigenvalue is here called λ to distinguish it from the analytical one (ζ). Also, the numerical eigenvector is indicated with $\{x\}$ to discern it from the analytical eigenfunction y. The numerical eigenvector is assumed to be normalized with respect to the generalized mass:

$$\{x\}^T[M]\{x\} = 1. \tag{9}$$

In a FE structural problem, $[K]$ is the stiffness matrix and $[M]$ can be either of the mass or differential stiffness matrix, based on the type of problem considered. The differentiation of this equation is due to Nelson [15,16] and the following equation is obtained:

$$\lambda' = \{x\}^T([K'] - \lambda[M'])\{x\}. \tag{10}$$

The method requires the derivatives of the stiffness and mass (or differential stiffness) matrices. Because of that, such an approach has not been applied to shape sensitivity problems until now. However, a nonintrusive and element agnostic approach has been developed in this work to calculate the derivative of structural matrices based on the primary analysis matrices and the connectivity of the mesh.

Applications and results
Since some sensitivity methods may not work with repeated eigenvalues, both approaches have been here applied to a vibration problem involving repeated eigenvalues. A beam with a circular cross-section ($r = 2.5\ mm$) and a length of $100\ mm$ has been considered. A Simply Supported-Sliding boundary condition has been applied in order to validate the method even when the stiffness matrix is singular. A uniform mesh containing forty beam elements has been created. The length of the beam has been considered as a shape design variable and a uniform design velocity has been employed. The reference values have been found employing the NASTRAN Design Sensitivity and Optimization solution (SOL 200). The comparison of the eigenvalue derivative with respect to the reference one for both CSA and Nelson's method is summarized in Table 1.

Table 1: First ten elastic natural frequencies and their derivative: comparison between SOL 200, CSA and improved Nelson's method.

Mode ID	Natural frequency [Hz]	SOL 200 Derivative [Hz]	CSA Derivative [Hz/mm]	Nelson's derivative [Hz/mm]
1,2	308.3	-6.1685	-6.1667 (-0.029 %)	-6.1664 (-0.034 %)
3,4	998.5	-19.972	-19.971 (-0.005 %)	-19.971 (-0.005 %)
5,6	2082	-41.642	-41.639 (-0.007 %)	-41.640 (-0.005 %)
7,8	3558	-71.162	-71.156 (-0.008 %)	-71.161 (-0.001 %)
9,10	5426	-108.52	-108.50 (-0.018 %)	-108.52 (-)

Both methods work with shape sensitivity problems and demonstrate very good accuracy. However, the CSA precision decreases if coarse meshes are employed. This is probably due to the many special gradients necessary for applying Eq. 7. They must be calculated numerically and their accuracy affects the final eigenvalue derivative estimation and also decreases the computational efficiency. Instead, Nelson's method is accurate even with coarse meshes and has a higher computational efficiency.

The two approaches have been then applied to buckling problems. A $1\ m$ long beam with a rectangular cross-section ($8\ mm \times 12\ mm$) has been here considered. A uniform mesh with twenty beam elements has been created. Several boundary conditions have been employed for the analysis: Simply Supported - Simply Supported (S-S), Clamped - Clamped (C-C), Clamped - Free

(C-F), Clamped - Guided (C-G), and Simply Supported - Guided (S-G). The buckling eigenvalues along with their derivatives are summarized in Table 2. The CSA and improved Nelson's method have been compared with the analytical results.

Table 2: Buckling eigenvalues and their derivative with respect to the beam length. CSA and enhanced Nelson's method comparison with analytical results.

Boundary Conditions	Eigenvalue [N]	Analytical derivative [N/mm]	CSA derivative [N/mm]	Nelson's derivative [N/mm]
S-S	359.29	-0.71857	-0.72480 (0.867 %)	-0.71857 (-)
C-C	1437.16	-2.87428	-3.02699 (5.313 %)	-2.87423 (-0.002 %)
C-F	89.821	-0.17964	-0.18011 (0.262 %)	-0.17963 (-0.006 %)
C-G	359.29	-0.71857	-0.72761 (1.258 %)	-0.71857 (-)
S-G	89.821	-0.17964	-0.18011 (0.262 %)	-0.17963 (-0.006 %)

Nelson's results perfectly match the analytical derivative, while CSA exhibits accuracy limitations even with enough fine meshes. In fact, when particularizing Eq. 7 to buckling problems, even more spatial gradients than vibration problems are required. As a result, Nelson's method is strongly suggested for this kind of application.

Conclusions
This work presented alternative methods to calculate shape design derivatives of beam eigenvalue problems. Two innovative solutions have been developed and investigated: the CSA and the enhanced Nelson's method. Both approaches have been successfully applied and validated. However, the CSA exhibited some limitations, especially in the buckling case. Nelson's method, on the contrary, has shown excellent accuracy, and very good computational efficiency. The enhanced Nelson's method can be successfully used in shape sensitivity problems and integrated into design optimization software. Future works will show the application of both approaches to plate and three-dimensional FE models.

References

[1] J. Iott, R.T. Haftka, H.M. Adelman, Selecting step sizes in sensitivity analysis by finite differences, National Aeronautics and Space Administration (NASA). NASA-TM-86382 (1985)

[2] P.E. Gill, W. Murray, M.H. Wright, Practical optimization, SIAM -Society for Industrial and Applied Mathematics, Philadelphia, 2019. ISBN: 978-1-61197-559-8

[3] R.T. Haftka, H.M. Adelman, Recent developments in structural sensitivity analysis, Structural optimization. 1 (1989) 137-151. https://doi.org/10.1007/BF01637334

[4] L.B. Rall, G.F. Corliss, Computational Differentiation: Techniques, Applications, and Tools, SIAM - Society for Industrial and Applied Mathematics. 89 (1996) 1-18.

[5] C.C. Margossian, A review of automatic differentiation and its efficient implementation, Wiley interdisciplinary reviews: data mining and knowledge discovery. 9 (2019). https://doi.org/10.1002/widm.1305

[6] J.R.R.A. Martins, P. Sturdza, J.J. Alonso, The complex-step derivative approximation, ACM Transactions on Mathematical Software (TOMS). 29 (2003) 245-262. https://doi.org/10.1145/838250.838251

[7] K. Dems, Z. Mróz, Variational approach by means of adjoint systems to structural optimization and sensitivity analysis - II: Structure shape variation, International Journal of

Solids and Structures, Elsevier. 20 (1984) 527-552. https://doi.org/10.1016/0020-7683(84)90026-X

[8] K. Dems, R.T. Haftka, Two approaches to sensitivity analysis for shape variation of structures, Mechanics of Structures and Machines, Taylor & Francis. 16 (1988) 501-522. https://doi.org/10.1080/08905458808960274

[9] K.K. Choi, N.H. Kim, Structural sensitivity analysis and optimization 1: linear systems, Mechanical Engineering Series, Springer, New York, 2004. https://doi.org/10.1007/b138709

[10] J. Borggaard, J. Burns, A PDE sensitivity equation method for optimal aerodynamic design, Journal of Computational Physics, Elsevier. 50 (1997) 366-384. https://doi.org/10.1006/jcph.1997.5743

[11] R. Duvigneau, D. Pelletier, On accurate boundary conditions for a shape sensitivity equation method, International journal for numerical methods in fluids, Wiley Online Library. 136 (2006) 147-164. https://doi.org/10.1002/fld.1048

[12] M.D. Kulkarni, R.A. Canfield, M.J. Patil, Nonintrusive continuum sensitivity analysis for fluid applications, Journal of Computational Physics, Elsevier. 403 (2020). https://doi.org/10.1016/j.jcp.2019.109066

[13] D.M. Cross, R.A. Canfield, Local continuum shape sensitivity with spatial gradient reconstruction, Structural and Multidisciplinary Optimization, Springer. 50 (2014) 975-1000. https://doi.org/10.1007/s00158-014-1092-0

[14] D.M. Cross, R.A. Canfield, Local continuum shape sensitivity with spatial gradient reconstruction for nonlinear analysis, Structural and Multidisciplinary Optimization, Springer. 51 (2015) 849-865. https://doi.org/10.1007/s00158-014-1178-8

[15] R.B. Nelson, Simplified calculation of eigenvector derivatives, AIAA journal. 14 (1976) 1201-1205. https://doi.org/10.2514/3.7211

[16] R.L. Dailey, Eigenvector derivatives with repeated eigenvalues, AIAA journal. 27 (1989) 486-491. https://doi.org/10.2514/3.10137

Aerospace Science and Engineering - IV Aerospace PhD-Days
Materials Research Proceedings 42 (2024) 43-46

Materials Research Forum LLC
https://doi.org/10.21741/9781644903193-10

Integrated process chain for crashworthiness assessment of innovative aircraft cabin layouts

Leonardo Marconi[1,a] *, Dieter Kohlgrueber[1,b], Michael Petsch[1,c]

[1]German Aerospace Center (DLR), Institute of Structures and Design (BT), Structural Integrity department, Pfaffenwaldring 38-40, 70569 Stuttgart, Germany

[a]leonardo.marconi@dlr.de, [b]dieter.kohlgrueber@dlr.de, [c]michael.petsch@dlr.de

Keywords: Process Chain, FEM, ATDs, Crashworthiness, Occupant Safety, CPACS

Abstract. Over the years, the German Aerospace Center (DLR) has developed a multidisciplinary process chain for aircraft design. This paper introduces a novel feature of this process chain, implemented in the Python tool called "Parametric Numerical Design and Optimization Routines for Aircraft" (PANDORA) developed at the Institute of Structures and Design (BT) featuring enhanced numerical tools for the fuselage structure analysis and sizing. The ongoing development focuses on achieving a high level of detail in the aircraft Finite Element (FE) modeling chain for crashworthiness evaluations, including a more detailed cabin description as well as advanced anthropomorphic test devices (ATDs). The aim is to study the dynamic behavior of novel cabin layouts and structures under crash conditions starting from their parametrical definition, and evaluate innovative design choices in terms of safety. Such a fully integrated process chain becomes necessary due to recent changes in aviation crashworthiness regulations, where the shift from prescriptive to performance-based requirements may significantly influence both the aircraft design and certification process.

Introduction

Crashworthiness in aviation is gaining increasing importance, as civil aviation is one of the fastest-growing sectors worldwide. In addition, the recent surge of interest in sustainable aviation paves the way for novel disruptive designs that may implicate an increased passenger density in the cabin, involving new safety problems to be explored. Since regulations prescribe that the structure must be designed to reduce loads acting on the occupants due to a crash to a tolerable level, the crash load mitigation will influence the overall design of a transport aircraft fuselage structure and hence it should be considered in an early phase of the design process [1].

Traditionally, the certification of seats and cabin is mainly considered independent of the structural behavior of the aircraft. In fact, the certification process requires the application of a standardized acceleration pulse directly to the seat structure. Nevertheless, statistics show a high influence of the aircraft structure on crash survivability [2]. From 2017, the certification authorities are starting to change the regulation to allow considering the combined crash performance of seats mounted to the airframe, rather than considering them separately [3].

As the aforementioned combined consideration in fully instrumented full-scale testing is prohibitive in terms of complexity and costs, extensive use of validated simulation techniques and the development of new modeling methodologies is of primary interest.

A Multidisciplinary Aircraft Design Framework

Plenty of subjects are involved in the design process of a new aircraft, such as aerodynamics, structural design, cabin concept, flight mechanics, costs and many others. This results in a complex Multidisciplinary Design Optimization problem demanding an efficient and accurate process chain to aim for the best pre-design solution and avoid expensive modifications at a later design stage.

Aerospace Science and Engineering - III Aerospace PhD-Days   Materials Research Forum LLC
Materials Research Proceedings 42 (2024) 43-46   https://doi.org/10.21741/9781644903193-10

An increasingly sophisticated aircraft predesign process chain was established as a result of the aggregation of the unique expertise of various DLR institutes specialized in different subjects. Such a collaborative process employs the Common Parametric Aircraft Configuration Schema (CPACS) data format [4] as a means of exchanging the aircraft description and analysis results.

The hierarchical CPACS data structure enables integration across disciplines and different levels of detail, ranging from simple statistical models to high-fidelity methods like FE analysis and computational fluid dynamics.

Within the DLR multidisciplinary process, PANDORA (Fig. 1) is an established tool for the automatic generation of global and detailed FE structural models for static and dynamic analysis (crash and ditching) and subsequent structural sizing [5]. A persistent effort is made to expand the modeling chain and achieve a higher level of detail. This paper introduces the new feature to integrate a realistic cabin environment (seats, luggage compartments, side panels, etc.) as well as ATDs into the FE model, with the final aim of evaluating under realistic crash scenarios.

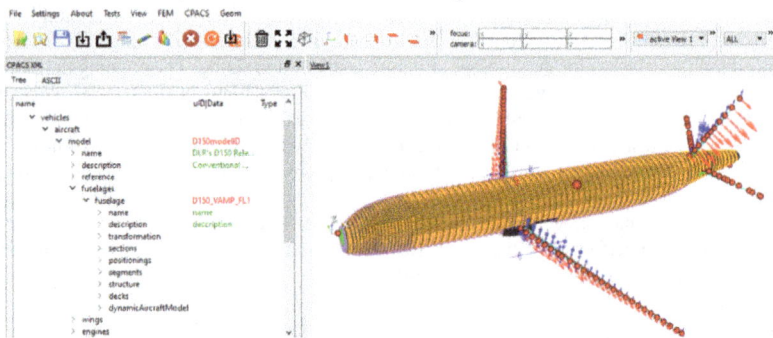

Figure 1. PANDORA graphical user interface. Visualization in PANDORA of CPACS data and resulting aircraft Global FEM

Crash Analysis of FE Parametric Models

The finite element model is automatically generated based on geometric and structural data in a CPACS file and stored in the general FE format of PANDORA. Pre-processing is possible directly in PANDORA through plenty of dedicated functions to create models in a wide range from low complexity to high fidelity. Within this process, the portion of fuselage to be analyzed can be selected, the mesh can be locally refined and each structural profile cross-section can be extruded in arbitrary zones, to achieve both, an accurate and cost-effective simulation.

After the model generation, the internal FE database is converted to the selected solver's format, according to the type of application. Currently, PANDORA features interfaces to standard structural solvers such as ANSYS, and NASTRAN as well as the dynamic solvers VPS and LS-DYNA with an explicit time integration schema. The integration of the passenger dummies shall be conducted in LS-DYNA due to its availability of dummy models representing a wide range of passengers, from light (5th percentile female) to heavy people (95th percentile male).

Two main topics were addressed prior to the integration of the dummies. The first topic (left side in Fig. 2) concerns the new capability of PANDORA to create LS-DYNA models. The validation of LS-DYNA modeling methodology was carried out with a bottom-up approach, though several numerical benchmarks, starting from simple static models up to the fuselage barrel crash. The results are compared with validated test runs in VPS, with its well-established modeling strategy previously used for ditching studies [6].

The second topic (right side in Fig. 2), involves the development of the methodology for modeling ATDs, belts, and seats. This is defined and validated in a parallel study [7]. A first simple seat model originating from studies performed in the 1990s is used in this first part of the project, but shall be replaced in later stages of the research project.

Currently, the integration of the dummy in the seat requires specialized manual work. Hence, to fit ATDs into an automatic process chain, a library single seat-ATD assembly is created for different ATD types, using the previously developed methodology. The single-seat-ATD models are then combined to simulate the different specific seat configurations defined in CPACS.

The automatic generation of a full-integrated cabin is sought as a result of this implementation (top image in Fig. 2, manually implemented previous department's model [8]). This will allow the creation of full-scale aircraft models in a matter of seconds, rather than days of manual work.

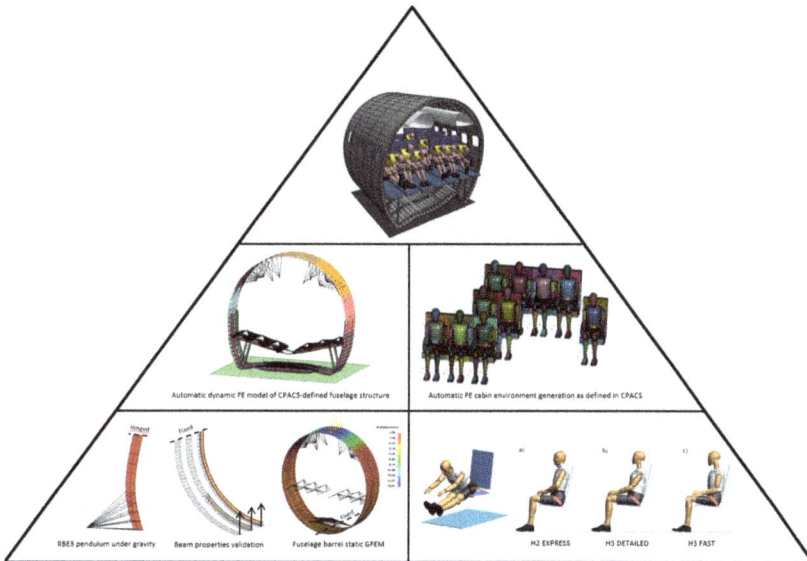

Figure 2. Integration pyramid in three steps from simple benchmarks to full-cabin model

Conclusion and Future Works

The DLR Institute of Structures and Design has extensive experience in creating FE models for aircraft crash analysis, which however requires a long integration work, with limited possibilities to modify the model. The knowledge of the institute in aircraft and ATD modeling is condensed into a novel PANDORA implementation, which paves the way to fully parametrized crashworthiness studies of full-scale aircrafts. This new capability represents a powerful tool to understand the outcomes of pioneering designs in terms of occupant safety, for innovative cabin layouts (Fig. 3), seat configurations, and their connection to the structure (Fig. 4).

This shall accelerate the development of innovative aircraft concepts, while aiding designers to improve aircraft safety, by considering crashworthiness from the first stages of the predesign phase. In particular, the emphasis is on addressing the safety challenges of emerging designs such as zero-emission configurations and high-density cabin layouts in which the design space is yet to explore and a means for parametric analysis is of fundamental importance.

Future developments of PANDORA include the integration of advanced ATDs and human body models, and the investigation of up-to-date seat structures to include in the structural model.

Figure 3. *Examples of novel cabin design and zero-emission aircrafts developed by DLR Institute of System Architectures in Aeronautics (SL)*

Figure 4. *Reconfigurable (left DLR SL) and high-density seat configurations (right [9])*

References

[1] M. Waimer, Development of a Kinematics Model for the Assessment of Global Crash Scenarios of a Composite Transport Aircraft Fuselage. DLR-Forschungsbericht, dissertation Universität Stuttgart. 274 S. (2013). https://elib.dlr.de/88358/

[2] R. G. Thomson, C. Caiafa: Structural response of transport airplanes in crash situations, report NASA TM85654, (1983). https://ntrs.nasa.gov/citations/19830019709

[3] EASA CS-23 Amendment 5

[4] M. Alder, E. Moerland, J. Jepsen, B. Nagel, Recent Advances in Establishing a Common Language for Aircraft Design with CPACS, Aerospace Europe Conference 2020. (2020). https://elib.dlr.de/134341/

[5] M. Petsch, D. Kohlgrüber, J. Heubischl, PANDORA - A python based framework for modelling and structural sizing of transport aircraft, MATEC Web of Conf. 233 00013 (2018). https://doi.org/10.1051/matecconf/201823300013

[6] C. Leon Muñoz, E. Wegener, M. Petsch, D. Kohlgrüber, Detailed FE aircraft fuselage sections for water impact simulations in the pre-design process chain, Journal of Physics: Conference Series 2526 (2023). https://iopscience.iop.org/issue/1742-6596/2526/1

[7] N. Wegener, C. Sauer, P. Schatrow M. Waimer, A Systematic Approach Towards Integrated Safety Modelling for Aerospace Applications - Preliminary Results on Rigid Seat Simulations, 14th European LS-DYNA Conference (2023). https://elib.dlr.de/200575/

[8] P. Schatrow, M. Waimer, M. Petsch, C. Leon Muñoz, D. Kohlgrüber, Method development for full aircraft crash simulation at different levels of modeling detail. The Ninth Triennial International Fire & Cabin Safety Research Conference. (2019). https://elib.dlr.de/130176/

[9] Image from https://www.dailymail.co.uk/travel/travel_news/article-5625283/The-standing-seats-thatll-let-airlines-cram-passengers.html

Aerospace Science and Engineering - IV Aerospace PhD-Days
Materials Research Proceedings 42 (2024) 47-50

Materials Research Forum LLC
https://doi.org/10.21741/9781644903193-11

Reading, transmission, storage, and display of near real-time data acquired by fiber Bragg grating sensors installed on an unmanned aerial vehicle

Antonio Costantino Marceddu[1,a,*], Bartolomeo Montrucchio[1,b]

[1]Politecnico di Torino, Corso Duca degli Abruzzi 24, 10129 Torino, ITALY

[a]antonio.marceddu@polito.it, [b]bartolomeo.montrucchio@polito.it

Keywords: Aerospace, Computer Graphics, Data Visualization, Database, Fiber Bragg Grating, Graphical User Interface, Middleware, Optical Fiber, Remotely Piloted Aircraft, Wireless Communication

Abstract. Fiber Bragg grating (FBG) sensors are a particular type of optical sensor capable of measuring multiple physical properties, like strain, temperature, and vibration. Their properties are of great interest for several use cases, including telecommunications, security, medicine, and biotechnology. This paper will discuss the progress of a project that uses them to monitor the health of an instrumented Unmanned Aerial Vehicle (UAV). Such operation is carried out on the ground in near real-time using software for intuitive data visualization.

Introduction

Optical fibers are thin, flexible strands of glass or clear plastic used to transmit light signals over long distances with minimal loss of signal strength. They have a layered structure:

- In the innermost part is the core, where the light signal travels.
- Next is the cladding, which has a composition that maximizes internal reflection and prevents light from escaping from the core so it can travel through the fiber.
- Finally, there are one or more polymer coatings that have a protection function from environmental and mechanical agents.

Inside the core, it is possible to carry out a periodic modulation of the refractive index. These changes modify the light passing through the fiber, causing it to reflect a specific wavelength of light called the Bragg wavelength. This microstructure is called Fiber Bragg Grating (FBG) and can be used as a sensor to measure various engineering parameters such as temperature, strain, vibration, pressure, and load. Fig. 1 depicts the FBG operation principle. The advantages optical fibers offer, such as lightweight and small size, immunity to electromagnetic interference, and reliability, make FBG sensors suitable for different applications, such as Structural Integrity Monitoring (SHM), medical sensing and imaging, environmental monitoring, and more. These applications are then widely used in geodynamics, telecommunications, civil structural engineering, the energy industry, and aerospace [1-11].

This paper precisely concerns this last example, as it deals with a research project concerning the use of FBG sensors for the structural monitoring of an Unmanned Aerial Vehicle (UAV). It will be discussed in detail in the next section.

Aerospace Science and Engineering - III Aerospace PhD-Days
Materials Research Proceedings 42 (2024) 47-50

Materials Research Forum LLC
https://doi.org/10.21741/9781644903193-11

Flying test bench

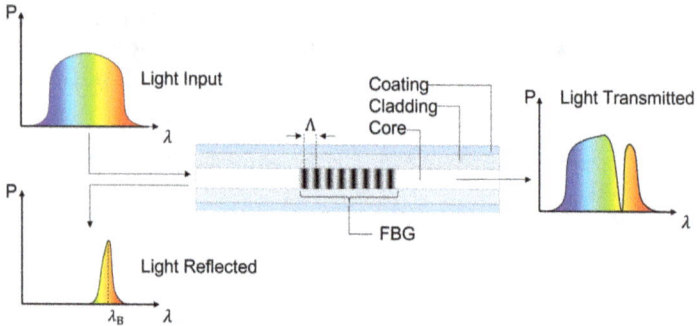

Figure 1 – An illustration of FBG working principle. When the light signal is transmitted through the fiber, a part of it, associated with the Bragg wavelength λ_B, is reflected back towards the source. The image was taken from [1].

The research project discussed in this paper is related to the use of optical sensors for an aerospace application. It was born as an internal collaboration at the PhotoNext Interdepartmental Center, a hub for applied optics of the Politecnico di Torino created in the summer of 2017 as a cornerstone of its strategic research initiatives. In detail, it involves the Department of Control and Computer Engineering (DAUIN) and the Department of Mechanical and Aerospace Engineering (DIMEAS) in the creation of a flying test bench to experiment with the use of FBG sensors in the aerospace sector and monitor the related values remotely [12-15]. The flying test bench, whose operating diagram is shown in Fig. 2, mainly consists of a UAV called Anubi. It was constructed in 2017 by the ICARUS student team of the Politecnico di Torino to participate in the Air Cargo Challenge (ACC) 2017, an important biennial European student competition for the field of aeronautical engineering. FBG sensors have been strategically placed on the wing, fuselage, and tail of the Anubi UAV to monitor its structure in real-time while in flight, providing valuable data for structural integrity assessment. The measured values can be read using an optical interrogator, to which the FBG sensors themselves must be connected. In the specific case in question, the optical interrogator used is the SmartScan produced by Smart Fibres Ltd. The remote monitoring operation can be made through a dedicated software application chain, which allows reading, transmitting, storing, and displaying the FBG sensor values directly on the ground via Personal Computers (PC). The Middleware represents the first element in this chain, an application written in C/C++ made to run directly on a Raspberry Pi mounted on the UAV itself and connected to the SmartScan via Ethernet. It is responsible for reading the data coming from the optical interrogator and sending it to a NoSQL Cloud Database based on MongoDB. One of the key advantages of NoSQL databases, which store data differently than relational tables, is that they are designed to provide fast read and write operations, making them suitable for use cases that require low latency and real-time data processing. Among the features of MongoDB is Change Stream, a real-time data update notification feature that allows applications that interface with the database to respond instantly to data changes. The last element of the chain is represented by the Viewers, whose primary purpose is to guarantee a simple and intuitive vision of the sensory data by researchers working on the type of optical sensors under examination. At the moment, there are two different Viewers, which allows the intuitive displaying of sensory data directly on the ground.

Materials Research Forum LLC
https://doi.org/10.21741/9781644903193-11

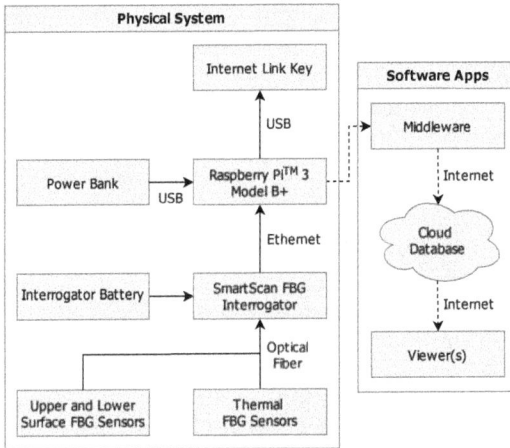

Figure 2 – A diagram depicting the operation of the flying test bench.

The PhotoNext 3D Viewer, realized with Unity, allows to display data in two ways [12]:

- A 3D heat map view offers a qualitative depiction of sensor values directly on a 3D model of the UAV. This is done by adjusting the color gradients of the 3D model near the area where each sensor is located.
- A graphical view shows the fluctuation of sensor values from their baseline.

PhotoNext FBG Data Analyzer was created without an actual game engine to make it more performant [15]. It has, therefore, lost the ability to view data via 3D heat map, but it is rich in multiple features and views, including tabular visualization of individual values, that are not present in PhotoNext 3D Viewer.

Conclusions

This paper discusses the current state of implementation of the flying test bench. Its purposes are multiple: to test the use of FBG sensors for the aerospace sector and to implement a remote monitoring system. So far, three flight test campaigns have been successfully conducted at the Tetti Neirotti runway, near Turin, to test both new FBG sensors and new applications. A new flight campaign is currently planned to test novel sensors.

Funding

This work was carried out under the PhotoNext initiative at Politecnico di Torino (http://www.photonext.polito.it/).

References

[1] F. Falcetelli, A. Martini, R. Di Sante, and M. Troncossi, Strain Modal Testing with Fiber Bragg Gratings for Automotive Applications, *Sensors* 22(3) (2022), 946. https://doi.org/10.3390/s22030946

[2] C.-Y. Hong, Y.-F. Zhang, M.-X. Zhang, L. M. G. Leung, and L.-Q. Liu, Application of FBG Sensors for Geotechnical Health Monitoring, a Review of Sensor Design, Implementation Methods and Packaging Techniques, Sensors and Actuators A: Physical, 244 (2016), 184-197, https://doi.org/10.1016/j.sna.2016.04.033.

[3] M. A. Riza, Y. I. Go, S. W. Harun, and R. R. J. Maier, "FBG Sensors for Environmental and Biochemical Applications—A Review," *IEEE Sensors Journal* 20(14) (2020), 7614-7627. https://doi.org/10.1109/JSEN.2020.2982446

[4] J. Leng and A. Asundi, Structural Health Monitoring of Smart Composite Materials by Using EFPI and FBG Sensors, Sensors and Actuators A: Physical, 103(3) (2003), 330-340. https://doi.org/10.1016/S0924-4247(02)00429-6

[5] C. Rodrigues, C. Félix, A. Lage, and J. Figueiras, Development of a Long-Term Monitoring System based on FBG Sensors Applied to Concrete Bridges, Engineering Structures 32(8) (2010), 1993-2002. https://doi.org/10.1016/j.engstruct.2010.02.033

[6] M. Mieloszyk and W. Ostachowicz, An application of Structural Health Monitoring system based on FBG sensors to offshore wind turbine support structure model, Marine Structures 51 (2017), 65-86. https://doi.org/10.1016/j.marstruc.2016.10.006

[7] X. Qiao, Z. Shao, W. Bao, and Q. Rong, Fiber Bragg Grating Sensors for the Oil Industry, *Sensors 17*(3) (2017), 429. https://doi.org/10.3390/s17030429

[8] O. De La Torre, I. Floris, S. Sales, and X. Escaler, Fiber Bragg Grating Sensors for Underwater Vibration Measurement: Potential Hydropower Applications, Sensors 21(13) (2021), 4272. https://doi.org/10.3390/s21134272

[9] A. Aimasso, C. G. Ferro, M. Bertone, M. D. L. Dalla Vedova, and P. Maggiore, Fiber Bragg Grating Sensor Networks Enhance the In Situ Real-Time Monitoring Capabilities of MLI Thermal Blankets for Space Applications, Micromachines 14(5) (2023), 926. https://doi.org/10.3390/mi14050926

[10] A. Aimasso, Optical fiber sensor fusion for aerospace systems lifecycle management, Materials Research Proceedings 33 (2023), 288-293. https://doi.org/10.21741/9781644902677-42

[11] A. Aimasso, G. Charruaz, M. Bertone, C. G. Ferro, M. D. L. Dalla Vedova, and P. Maggiore, Test Bench and Control Logic Development for Dynamic Thermal Characterization of Optical Sensors, International Journal of Mechanics and Control 24 (2023).

[12] A. C. Marceddu, G. Quattrocchi, A. Aimasso, E. Giusto, L. Baldo, M. G. Vakili, M. D. L. Dalla Vedova, B. Montrucchio, and P. Maggiore, Air-to-Ground Transmission and Near Real-Time Visualization of FBG Sensor Data via Cloud Database, *IEEE Sensors Journal* 23(2) (2023), 1613-1622. https://doi.org/10.1109/JSEN.2022.3227463

[13] A. C. Marceddu, A. Aimasso, A. Scaldaferri, P. Maggiore, B. Montrucchio, and M. D. L. Dalla Vedova, Creation of a Support Software for the Development of a System for Sending and Visualizing FBG Sensor Data for Aerospace Application, *2023 IEEE 10th International Workshop on Metrology for AeroSpace (MetroAeroSpace)* (2023) 487-491. https://doi.org/10.1109/MetroAeroSpace57412.2023.10190008

[14] A. C. Marceddu and B. Montrucchio, Storage and visualization on-the-ground and in near real-time of the data measured by the optical sensors connected to a flying test bench, Materials Research Proceedings 33 (2023), 277-280.

[15] A. C. Marceddu, A. Aimasso, S. Schiavello, B. Montrucchio, P. Maggiore, M. D. L. Dalla Vedova, Comprehensive Visualization of Data Generated by Fiber Bragg Grating Sensors in *IEEE Access* 11 (2023). doi: https://doi.org/10.1109/ACCESS.2023.3329425

Aerospace Science and Engineering - IV Aerospace PhD-Days
Materials Research Proceedings 42 (2024) 51-55

Materials Research Forum LLC
https://doi.org/10.21741/9781644903193-12

Least-weight design of tow-steered composites

Dario Zamani[1,a] *

[1]Politecnico di Torino, Corso Duca degli Abruzzi, 10129, Turin, Italy

[a]dario.zamani@polito.it

Keywords: Variable Stiffness Composites, Unified Formulation, Discrete Optimization, Least-Weight

Abstract. Automatic Fiber Placement (AFP) enables spatial variation of fiber orientation, enhancing mechanical performance compared to traditional composites. Variable Angle Tow (VAT) or Variable Stiffness Composites (VSC) optimize structural efficiency, which is crucial for lightweight aerospace structures. However, it is important to consider that limitations resulting from the manufacturing process can significantly impact the design domain available. This study proposes a mixed-integer optimization approach integrating the Carrera Unified Formulation (CUF) to minimize laminate weight while meeting frequency performance. This research aims to determine the optimal number of layers and lamination angles, considering manufacturing constraints and evaluating the impact of the selection of structural theory on the solutions.

Introduction

In contrast to conventional straight fiber laminates, Variable Stiffness Composites (VSCs) provide designers with a notably expanded design space by enabling control over the fiber tows along conforming curvilinear paths. However, due to the inherently non-uniform stiffness properties of the computational domain, resulting in increased computational costs, it is imperative to develop suitable and efficient optimization and design tools. Groh and Weaver [1] addressed the issue of reducing the mass of VAT laminates produced by Continuous Tow Shearing (CTS). In particular, a genetic algorithm (GA) was combined with pattern searching, creating a hybrid optimization approach that reduced weight by 31% compared to traditional straight-fiber composites. Catapano et al. [2] proposed a two-level optimization strategy for VAT laminates that considers manufacturing requirements for Fused Filament Fabrication (FFF) and Continuous Filament Fabrication (CFF) techniques. The strategy involved multi-scale optimization and emphasized incorporating technological constraints into the design optimization process. In addition, Sánchez-Majano and Pagani [3] implemented the Carrera Unified Formulation (CUF) to construct a surrogate that mimicked the objective function to maximize the buckling load and the fundamental frequency of the VAT plates. This method provided the advantage of obtaining a model whose accuracy was determined by the appropriate choice of the order of the structural theory adopted. Lastly, Pagani et al. [4] developed a surrogate optimization framework to maximize the first natural frequency of VAT laminates, taking into account the presence of gaps and overlaps by combining the CUF with the Defect Layer Method (DLM) [5] directly into the optimization process.

This work aims to present a mixed-integer optimization strategy specifically designed to select the least-weight design of a VAT laminate while also maximizing the first natural frequency. The research has two main objectives: to determine how the manufacturing constraints affect the minimum number of layers needed to meet fundamental frequency requirements and to examine how the selection of structural theory affects optimal solutions.

Unified finite elements for VSCs

In this work, CUF formalism is used to implement 2D FE. Specifically, as stated in [6], the 3D displacement field can be expressed using arbitrary through-the-thickness expansion functions $F_\tau(z)$ as follows:

$$u(x,y,z) = F_\tau(z)u_\tau(x,y). \ \tau = 1, ..., M \tag{1}$$

The symbol M represents the number of expansion terms, and $u_\tau(x,y)$ represents the vector containing the generalized displacements, with τ indicating summation. Examining multi-layered structures typically employs either an Equivalent-Single-Layer (ESL) or Layer-Wise (LW) approach. In this manuscript, ESL models utilize Taylor polynomials represented by $F_\tau(z)$; while LW employs Lagrange polynomials over individual layers and ensures the continuity of displacements at layer interfaces. Utilizing the FE and shape functions $N_i(x,y)$, the displacement field becomes:

$$u(x,y,z) = N_i(x,y)F_\tau(z)q_{\tau i}(x,y). \ i = 1, ..., N_n \tag{2}$$

In Eq. (2), $q_{\tau i}$ are the unknown nodal variables, with N_n denoting the number of nodes per element. This work employs 2D nine-node quadratic elements for the in plane discretization.

The governing equations are derived using the Principle of Virtual Displacements (PVD), which asserts that the virtual variation of internal strain energy, $\delta\mathcal{L}_{int}$, equals the virtual work of external forces, $\delta\mathcal{L}_{ext}$, minus inertia forces, $\delta\mathcal{L}_{ine}$. Specifically, $\delta\mathcal{L}_{int}$ can be written as:

$$\delta\mathcal{L}_{int} = \delta q_{sj}^T \left[\int_V D^T(N_j F_s) \, \tilde{C} D(N_i F_\tau) \, dV \right] q_{\tau i} = \delta q_{sj}^T k_0^{ij\tau s} q_{\tau i}, \tag{3}$$

where $k_0^{ij\tau s}$ is the 3×3 Fundamental Nucleus (FN) of the stiffness matrix, invariant regardless of 2D shape function order and through-the-thickness expansion. $D(\cdot)$ is the differential operator matrix with geometric relations, and $\tilde{C}$ is the material stiffness matrix in the global reference frame, i.e., $\tilde{C} = T(x,y)^T C T(x,y)$. As fibers vary point-wise within the plane, the rotation matrix T also varies accordingly. The virtual work of the inertia forces can be expressed as:

$$\delta\mathcal{L}_{ine} = \delta q_{sj}^T \left[\int_V \rho I N_i N_j F_\tau F_s \, dV \right] \ddot{q}_{\tau i} = \delta q_{sj}^T m^{ij\tau s} \ddot{q}_{\tau i}, \tag{4}$$

in which I denotes the 3×3 identity matrix and $m^{ij\tau s}$ is the 3×3 FN of the mass matrix. Thus, the undamped free vibration problem is expressed as follows:

$$M\ddot{q} + Kq = 0. \tag{5}$$

In Eq. (8), M and K are the overall mass and stiffness matrices, respectively, obtained by iterating over FN's with indices i, j, τ and s to compute element-level matrices, which are then assembled for the global structure. By applying harmonic solutions $q = \tilde{q}e^{i\omega t}$, Eq. (8) becomes:

$$(K - \omega_i^2 M)\tilde{q}_i = 0 \tag{6}$$

where ω_i and $\tilde{q}_i$ are the ith natural frequency and eigenvector, respectively.

Optimization framework

This manuscript considers the multi-objective optimization of VSCs, which aims to minimize the weight and, therefore, the number of layers in the laminate while maximizing the first natural frequency. The plies that make up the laminate are varied for each evaluation of the objective function, along with the lamination angles T_0 and T_1 for each layer considered. Following the linearly varying fiber formulation [7], T_0 represents the fiber angle at the plate's center, while T_1 represents the angle at the edge. Both the unconstrained and the constrained problem are

Aerospace Science and Engineering - III Aerospace PhD-Days
Materials Research Proceedings 42 (2024) 51-55

Materials Research Forum LLC
https://doi.org/10.21741/9781644903193-12

considered. The maximum curvature of the AFP machine is chosen as the constraint of the optimization problem. Therefore, the local fiber curvature κ must be less than $\kappa_{lim} = 3.28 \, m^{-1}$, as seen in Eq. (7).

$$\kappa = \frac{2(T_1 - T_0)}{a} \cos\left((T_1 - T_0)\frac{x}{a/2} + T_0\right) \leq 3.28 m^{-1} \tag{7}$$

The NSGA-II algorithm is utilized to solve the optimization problem. The variables T_0 and T_1 can continuously vary between -90° and 90°, while the number of plies N_{ply} can discretely vary between 6 and 10 layers. At each iteration, input files for structural analysis are created, and the natural frequency is then evaluated using the CUF-based FE code, as described in Section 2. The process is repeated iteratively until convergence.

Results

This work aims to minimize the mass of a traditional 8-layer symmetrical straight fiber composite with optimum lamination θ = [0°, 90°, 0°, 90°]$_S$, found in [8], while also fulfilling first natural frequency performance. The width and length of the plate are $a = b = 0.5$ m, and each ply has a thickness of 0.159 mm. A fully clamped boundary condition was imposed on all four sides. The reference mass is $m_{ref} = 0.5247$ kg, while the optimum frequency is $f_{ref} = 54.68$ Hz. The convergence analysis performed in [4] suggests that a 6 x 6 Q9 FE mesh effectively captures the fundamental frequency, requiring relatively low computational effort, which facilitates the reduction of the computational burden associated with the optimization process.

Table 1. Optimal design results for the unconstrained problem

	ESL – TE 1	ESL – TE 3	LW – LD2
$\langle T_0, T_1 \rangle^1$ [°]	$\langle -90, 6 \rangle$	$\langle -90, 7 \rangle$	$\langle -90, 6 \rangle$
$\langle T_0, T_1 \rangle^2$ [°]	$\langle 90, -9 \rangle$	$\langle 90, -9 \rangle$	$\langle 90, -9 \rangle$
$\langle T_0, T_1 \rangle^3$ [°]	$\langle 90, -9 \rangle$	$\langle 90, -8 \rangle$	$\langle 90, -9 \rangle$
$\langle T_0, T_1 \rangle^4$ [°]	$\langle -90, 6 \rangle$	$\langle -90, 8 \rangle$	$\langle -90, 6 \rangle$
f_1 [Hz]	55.43$^{+1.37\%}$	54.99$^{+0.57\%}$	54.98$^{+0.55\%}$
mass [kg]	0.4591$^{-12.5\%}$	0.4591$^{-12.5\%}$	0.4591$^{-12.5\%}$

Table 1 presents the results of the multi-objective optimization for the unconstrained problem. As the structural theory is varied, the optimal lamination angles are shown. In addition, the implementation of VAT composites allows the use of a symmetrical 7-layer laminate, resulting in a mass reduction of 12.5% while maintaining approximately the same fundamental frequency as the reference model.

Table 2. Optimal design results for the constrained problem ($\kappa_{lim} = 3.28 \, m^{-1}$)

	ESL – TE 1	ESL – TE 3	LW – LD2
$\langle T_0, T_1 \rangle^1$ [°]	$\langle -86, -31 \rangle$	$\langle -85, -31 \rangle$	$\langle -85 - 30 \rangle$
$\langle T_0, T_1 \rangle^2$ [°]	$\langle 79, 26 \rangle$	$\langle 72, 22 \rangle$	$\langle 76, 25 \rangle$
$\langle T_0, T_1 \rangle^3$ [°]	$\langle -60, -12 \rangle$	$\langle -61, -13 \rangle$	$\langle -61, -13 \rangle$
$\langle T_0, T_1 \rangle^4$ [°]	$\langle 58, 10 \rangle$	$\langle 70, 20 \rangle$	$\langle 62, 15 \rangle$
f_1 [Hz]	52.42$^{-4.12\%}$	52.06$^{-4.79\%}$	52.01$^{-4.87\%}$
mass [kg]	0.4591$^{-12.5\%}$	0.4591$^{-12.5\%}$	0.4591$^{-12.5\%}$

As described in Section 3, Table 2 displays the optimal results for the constrained problem for both ESL and LW models. Similar to the previous case, a 12.5% reduction in mass is observed.

However, due to the significant decrease in the design domain resulting from the limitation of the maximum bending radius of the AFP machine, the natural frequency is reduced by slightly more than 4% with respect to the reference.

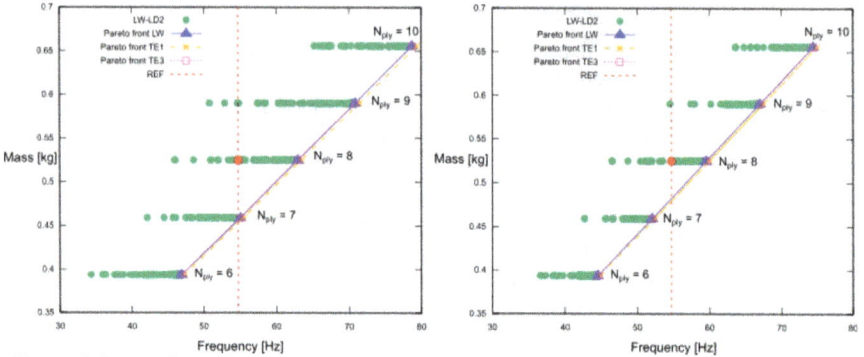

Figure 1. Pareto fronts for the unconstrained case (on the left) and for the constrained case (on the right)

Figure 1 displays the Pareto fronts for the unconstrained and constrained problem as the structural theory changes. It should be noted that the mixed-integer optimization strategy produces discontinuous response surfaces, allowing each layer to be optimized independently. In the unconstrained case, achieving the same optimum frequency as the reference utilizing a layer less is possible. However, in the constrained case, this is not possible. Nevertheless, it is noted that it is feasible to obtain much higher frequencies, but with the same mass as the reference

Conclusions

This work presented a mixed-integer optimization framework for the least-weight design of VSCs, taking into account the design space restriction resulting from the AFP manufacturing process. As demonstrated in [3], the structural theory exhibits a weak dependence on the optimal lamination angles for each layer. Moreover, layer-wise optimization provides discontinuous response surfaces, enabling the implementation of high-fidelity models that offer significant advantages over alternative optimization approaches.

References

[1] Groh, R.M., Weaver, P.: Mass optimisation of variable angle tow, variable thickness panels with static failure and buckling constraints. In: 56thAIAA/ ASCE/AHS/ASC Structures, Structural Dynamics, and Materials Conference (2015). https://doi.org/10.2514/6.2015-0452

[2] Catapano, A., Montemurro, M., Balcou, J. – A., Panettieri, E.,: Rapid prototyping of variable angle-tow composites. Aerotecnica Missili e Spazio 98(4), 257-271 (2019). https://doi.org/10.1007/s42496-019-00019-0

[3] Racionero Sánchez-Majano, A., Pagani, A.: Buckling and fundamental frequency optimization of tow-steered composites using layerwise structural models. AIAA Journal 61(9), 4149–4163 (2023). https://doi.org/10.2514/1.j062976

[4] Pagani, A., Racionero Sánchez-Majano, A., Zamani, D., Petrolo, M., Carrera, E.: Fundamental frequency layer-wise optimization of tow-steered composites considering gaps and overlaps. Aerotecnica Missili e Spazio (2024). (Under review)

[5] Fayazbakhsh, K., Nik, M. A., Pasini, D., Lessard, L.: Defect layer method to capture effect of gaps and overlaps in variable stiffness laminates made by automated fiber placement. Composite Structure 97 (2013), 245-251. https://doi.org/10.1016/j.compstruct.2012.10.031

[6] Carrera, E., Cinefra, M., Petrolo, M., Zappino, E.: Finite Element Analysis of Structures through Unified Formulation. Wiley & Sons, Hoboken, New Jersey. 2014.

[7] Gürdal, Z., Tatting, B.F., Wu, C.K.: Variable stiffness composite panels: Effects of stiffness variation on the in-plane and buckling response. Composites Part A: Applied Science and Manufacturing 39(5), 911–922 (2008). https://doi.org/10.1016/j.compositesa.2007.11.015

[8] Carvalho, J., Sohouli, A., Suleman, A.: Fundamental Frequency Optimization of Variable Angle Tow Laminates with Embedded Gap Defects. Journal of Composites Science 6, 64 (2022). https://doi.org/10.3390/jcs6020064

Aerospace Science and Engineering - IV Aerospace PhD-Days
Materials Research Proceedings 42 (2024) 56-60

Materials Research Forum LLC
https://doi.org/10.21741/9781644903193-13

Influence of fiber misalignment on the thermal buckling of variable angle tows laminated plates

Francesca Bracaglia[1,a] *

[1]DIMEAS, Mul2 Lab., Politecnico di Torino, Corso Duca degli Abruzzi 24, Torino, Italy

[a]francesca.bracaglia@polito.it

Keywords: Variable Angle Tow Composite, Misalignment Sensitivity, Thermal Buckling

Abstract. Variable Angle Tows (VAT) are a class of composite materials with curved fiber paths within the lamina plane. The tailoring of VATs is a virtue that increased the interest in their study. On the other hand, the same freedom in the fiber deposition inevitably leads to manufacturing defects like misalignment that impact the structure behavior, particularly buckling. The present study investigates the influence of manufacturing fiber misalignment on the thermal buckling response of a VAT composite square plate. The governing equations are obtained within the Carrera Unified Formulation (CUF) framework combined with Finite Element Method. The thermal problem is schematized with a decoupled approach, and the critical loads are evaluated through the solution of an eigenvalue problem. The results show how the presence of random misalignment influences both the buckling critical temperature and the buckling mode.

Introduction

The behavior of aerospace structures, particularly those constructed from composite materials, is highly influenced by the thermal environment due to various sources, such as solar radiation or drag in high-speed vehicles. Composites are appreciated in engineering due to their high stiffness and strength-to-weight ratios. They offer tailorable properties and are customizable by acting on stacking sequences or fiber deposition. The introduction of Variable Angle Tows (VAT) further enhances this tailoring capability, enabling curved fiber paths within lamina planes. During these years, the use of VAT has been extensively studied, particularly in the reinforcement around holes [1] and in the optimization of the deposition to retarding mechanical [2] and thermal buckling [3].

While VAT composites present opportunities for improved stiffness and strength properties, their utilization necessitates complex manufacturing techniques, leading to inevitable defects that have a significant impact on some specific phenomena. As a consequence, several studies have been published investigating their influence on failure mechanism [4], mechanical [5], and thermal buckling [6].

This work aims to evaluate the influence of misalignment due to manufacturing on the thermal buckling response of VAT plate, and random fields are employed to consider the random imperfections. The analyses are conducted through a buckling linearized formulation. The thermal problem is schematized using a decoupled approach where the thermal field is assumed to be known at each plate point and treated as an external load. The governing equations are obtained through the Principle of Virtual Displacements (PVD) within the Carrera Unified Formulation (CUF) framework [7] combined with the Finite Element Method (FEM). High-order theories are employed to evaluate the buckling response of the plate, allowing a better description of the results thanks to a more accurate thermal stress evaluation. The buckling results of a thousand analyses are collected, and some statistical conclusions are reported.

Thermal buckling problem

Through the CUF, it is possible to express the principal unknown of the problem which are the 3D displacements $u(x, y, z)$ by the combination of expansion functions $F_\tau(z)$ and 2D displacement components $u_\tau(x,y)$. Furthermore, the FEM is employed, and $u_\tau(x,y)$ can be expressed by combining the shape functions $N_i(x,y)$ and the nodal displacement vector $q_{i\tau}$ as represented in Eq. (1).

$$u(x,y,z) = F_\tau(z)\,u_\tau(x,y) = F_\tau(z)N_i(x,y)\,q_{i\tau} \quad \tau = 1,2,\dots,M \quad i = 1,2,\dots,N_n \qquad (1)$$

Where the double index means sum, M is the number of expansion terms, and N_n is the number of nodes. More details about the CUF and the allowable expansion functions can be found in [7].

The governing equations are obtained considering a decoupled thermal approach where strain and stresses are decomposed in pure mechanical and thermal components, as reported:

$$\varepsilon = \varepsilon_m + \varepsilon_\theta = b\,u - \alpha\,\Delta T \qquad (2)$$
$$\sigma = C\,\varepsilon = C(b\,u - \alpha\,\Delta T) = \sigma_m - \beta\Delta T \qquad (3)$$

Where β is a vector containing the terms coupling mechanical and thermal fields, and ΔT is the applied overtemperature. C is the material properties matrix and b is the geometrical operator. More details about these relations can be found in [5].

Through the use of a linearized approach, it is possible to find the thermal buckling critical load as the solution to the eigenvalue problem reported in Eq. (4).

$$\delta q^T : [\,K + \lambda_{cr}\,K_\sigma\,]\,\delta q = 0 \qquad (4)$$

Where K is the usual stiffness matrix and K_σ is the geometric stiffness matrix that is obtained from the expression of the variation of the work done by the virtual non-linear strains with the thermal initial stresses. The geometric stiffness matrix is assumed to be directly proportional to the buckling critical temperature λ_{cr}.

Misalignment through random field

The fiber path of VAT is described varying linearly according to the notation introduced by Gürdal et al. [8], the lamination is denoted as $[\Phi < T_0, T_1 >]$, where Φ is the reference system rotation angle from x-axis, T_0 and T_1 are the fiber orientation in the center and at the edge of the plate, respectively.

Fiber misalignments are modeled at the layer level through the stochastic field. They belong to a class of defects whose position is uncertain, so a statistical approach is necessary to introduce these defects into the analyses. Misalignment affects the local fiber path, and different orientations must be considered in the corresponding Gauss points.

In the present work, the Karhunen-Loève expansion (KLE) [9] is employed to describe the random fluctuation. Following this expansion, the stochastic field can be expressed as in Eq. (6).

$$\Delta H^k(x,y,\gamma) = \sum_{i=1}^{\infty} \zeta_i(\gamma)\sqrt{\lambda_i}\,\phi(x,y) \qquad (6)$$

Where H denotes the stochastic field and ΔH^k is the Gaussian variation of the random field at each layer. ζ_i denotes a set of random variables, λ_i and ϕ are the eigenvalues and the eigenvectors of the autocovariance kernel obtained solving the homogeneous Fredholom integral equation. Among the different solutions of the equation, the exponential function is here employed.

$$C(x,x',y,y') = \sigma_{H^k}^2 e^{-\frac{|x-x'|}{l_x} - \frac{|y-y'|}{l_y}} \qquad (7)$$

Where l_x and l_y are the correlation lengths in x and y direction and σ is the standard deviation of the random field.

Aerospace Science and Engineering - III Aerospace PhD-Days   Materials Research Forum LLC
Materials Research Proceedings 42 (2024) 56-60   https://doi.org/10.21741/9781644903193-13

Numerical results

The analysis is focused on a square plate composed of Kevlar/Epoxy with 150 mm edges length and 1.016 mm thickness. The plate is simply supported on all the edges and, one degree overtemperature along the entire plate is applied. Material properties are E_1 = 80 GPa, E_2 = 5.5 GPa, v=0.34, G_{12} = 2.2 GPa, G_{23} = 1.8 GPa, α_1= -0.9 x 10^{-6} 1/K, α_2= 27.0 x 10^{-6} 1/K. The lamination is $[\pm \theta]_s$ where θ is $[0 < 66.05, 11.73 >]$ corresponding to the optimum fiber deposition retarding thermal buckling [3].

Convergence analysis is conducted and 16 x 16 bilinear FEM elements discretization is chosen to ensure a good convergence of the results. Furthermore, a Lagrange Expansion of the second order LE2 is selected as expansion theory along the thickness.

The employed KLE has 15 parameters in the expansion for each layer and, the correlation lengths are l_x = 0.075, l_y=0.0075. The standard deviation of the random field is σ = 1.5 and the mean of the field is zero. Fig. 1 reports an example of the introduced random misalignment compared to the pristine fiber deposition for the first and second layers of the plate.

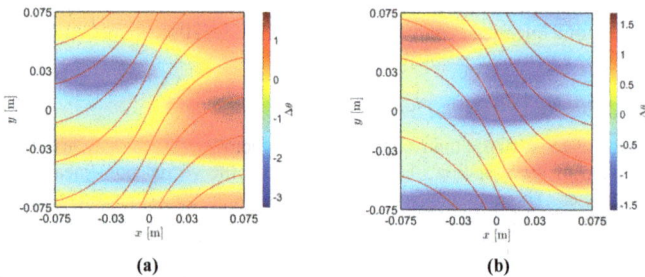

(a)   (b)

Figure 1: Misalignment due to Random Field and pristine fiber deposition. (a) First layer and, (b) Second layer

The first critical load of the pristine plate is equal to 22.02 °C, while the corresponding reference result is 22.28 °C [3], demonstrating the validity of the present method.

In order to investigate the influence of the misalignment on the critical thermal load, a spice of one thousand analyses is selected. The Probability Density Functions (PDFs) for the first five critical temperatures are reported in Fig.s 2(a) and 2(b), where it is clear that the first and second critical load PDFs are not overlapped and there is a considerable distance between the two critical temperature. The PDFs reported in Fig. 2(b) make clear that the third and fourth buckling loads are overlapped in a certain range of temperatures. To make it clear, Fig. 2(c) reports the number of repetition of the critical load λ_3 and, λ_4.

Fig. 3 shows the mean value and the standard deviation of the Modal Assurance Coefficient (MAC) of the buckling modes [5].

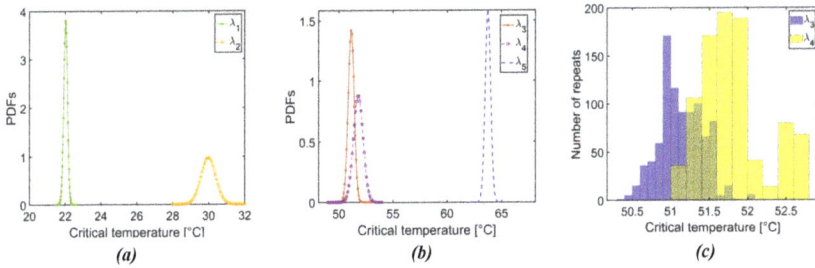

Figure 2: PDFs (a), (b), and number of repetition (c) of the critical temperature for the analyses of the spice.

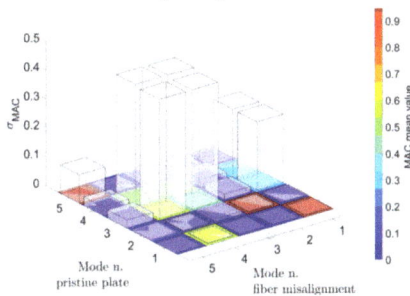

Figure 3: Mean value and standard deviation of the Modal Assurance Coefficient of the analyses.

The standard deviation and mean value of the MAC allow some considerations. The first two modes and the fifth are linearly independent of each other, and there are no interaction phenomena between them. On the other hand, the third and fourth modes present a switch governed by the presence of random misalignments. Therefore, the standard deviation corresponding to these modes is non-zero, and the switching is probably due to the symmetry of load and constraints. Finally, mode 4 presents a combination given by mode 1. This mode consists of a small central deflection with two larger internal deflections, which implies that this mode is not entirely independent of mode 1 that consisting of a single central deflection. The standard deviation of the MAC value is about zero, indicating that this result is present for about each analysis and is independent of the nature of the random misalignment.

Conclusions

The present work focuses on the evaluation of random fiber misalignment on the thermal buckling response of a simply supported square VAT plate composed of Kevlar/epoxy. From the numerical results, it is possible to say that the presence of misalignment does not highly influence the first and second buckling modes of the plate. On the other hand, the third and fourth modes present a switch to the buckling mode and an overlap of the critical temperatures. Future work can be addressed to the study of other materials, plates, or other classes of manufacturing defects.

References

[1] Y. Zhu, Y. Qin, S. Qi, H. Xu, D. Liu, C. Yan. Variable angle tow reinforcement design for locally reinforcing an open-hole composite plate. *Compos Struct*, 2018;202:162–9.

[2] R. Vescovini, V. Oliveri, D. Pizzi, L. Dozio, P.M. Weaver. Pre-buckling and Buckling Analysis of Variable-Stiffness, Curvilinearly Stiffened Panels. *Aerotec. Missili Spaz.*, 2020;99:43-52. https://doi.org/10.1007/s42496-019-00031-4

[3] A.V. Duran, N.A. Fasanella, V. Sundararaghavan, A.M. Waas. Thermal buckling of composite plates with spatial varying fiber orientations. *Compos. Struct.* (2015):124:228–35. https://doi.org/10.1016/j.compstruct.2014.12.065

[4] A. Pagani, and A.R. Sanchez-Majano. Stochastic stress analysis and failure onset of variable angle tow laminates affected by spatial fibre variations. *Compos. Part C,* 4 (2021): 100091. https://doi.org/10.1016/j.jcomc.2020.100091

[5] A.R. Sanchez-Majano, A. Pagani, M. Petrolo, C. Zhang. Buckling sensitivity of tow steered plates subjected to multiscale defects by high-order finite elements and polynomial chaos expansion. *Materials 2021*;14(11):2706. https://doi.org/10.3390/ma14112706

[6] N. Sharma, M. Nishad, D.K. Maiti, M.R. Sunny, B.N. Singh,. Uncertainty quantification in buckling strength of variable stiffness laminated composite plate under thermal loading. *Compos. Struct.* 275 (2021): 114486. https://doi.org/10.1016/j.compstruct.2021.114486

[7] E. Carrera, M. Cinefra, M. Petrolo, and E. Zappino. *Finite Element Analysis of Structures through Unified Formulation.* John Wiley & Sons, Chichester, West Sussex, UK, 2014.

[8] Z. Gürdal, B.F. Tatting, C.K. Wu. Variable stiffness composite panels: Effects of stiffness variation on the in-plane and buckling response. Composites Part A: Applied Science and Manufacturing, 39(5):911–922, May 2008.

[9] R.G. Ghanem, P.D. Spanos. *Stochastic finite elements: a spectral approach.* Dover Publications, Inc. 31 East 2nd Street, Mineola, N.Y. 11501

Aerospace Science and Engineering - IV Aerospace PhD-Days
Materials Research Proceedings 42 (2024) 61-65

Materials Research Forum LLC
https://doi.org/10.21741/9781644903193-14

2D higher-order theories for progressive damage model of composite structures based on Hashin and Puck failure criteria

Elisa Tortorelli[1,a] *, Salvatore Saputo[1,b] and Erasmo Carrera[1,c]

[1]Department of Mechanical and Aerospace Engineering, Politecnico di Torino, Turin, Italy, 10129

[a]elisa.tortorelli@polito.it, [b]salvatore.saputo@polito.it, [c]erasmo.carrera@polito.it

Keywords: CUF, Explicit Damage Modelling, Failure Criteria

Abstract. This paper proposes a high-order 2D finite element model for the progressive damage model of composite structures. The model is based on Carrera Unified Formulation (CUF), which allows to automatically implement different kinematics by using an opportune recursive notation. A Newton-Raphson algorithm and the explicit integration scheme is used to find the converged solution. A single element and an open-hole specimen under tensile and compression loads are investigated using a damage model based on Hashin and Puck failure criteria. The proposed model is compared with literature and ABAQUS continuum shell results.

Introduction

In recent decades, advanced fiber-reinforced composites have significantly integrated into various high-performance applications, including the aerospace, automotive, and maritime sectors. Nevertheless, composite materials are not immune from degradation, damage and nonlinear behavior. There are numerical tools for modeling progressive failure of composite across scales through development of computationally-efficient advanced structural models, as micromechanical progressive failure analysis [1].

Computational damage models for composite structures can be broadly categorized into two main approaches. The first approach, known as discrete modeling, entails the explicit geometrical representation of cracks within the structure, but it comes with significantly heightened computational demands. A different approach to discrete damage modeling involves continuum damage mechanics (CDM), which preserves the continuity of the displacement field within the finite element mesh [2] showing low computational cost. The fracture energy is used to mitigate a strong mesh dependency. A purely continuum damage approach is described in Ref. [3], focusing on intralaminar damage within the ply, which is characterized using the Composite Damage (CODAM) model. In particular, it is used the second-generation damage model, CODAM2 which is a strain-based damage formulation [4].

This paper consider a new damage model based on Hashin 3D failure criteria and Puck's failure criteria for matrix compression [6]. The model is compared with CODAM2 implemented by Nagaraj et al. [5], which is based on Hashin 2D failure criteria. The structural modelling is built in the framework of the Carrera Unified Formulation (CUF) [7] which allows employing higher-order structural theories to develop refined 1D and 2D models where the kinematic field is enriched via the use of cross-section and thickness expansion functions, respectively. This approach reduces the computation cost compared with standard solid elements FE analyses. The explicit integration scheme is used to find the converged solution.

Failure criteria and damage model

The CODAM2 material model implemented in the CUF framework has been previously explained in Ref. [5, 8], which consider Hashin 2D quadratic failure criterion for fiber and matrix tensile/compression. In this paper, instead, three-dimensional Hashin criteria are implemented for matrix, fiber tensile and fiber compression. For matrix compression Puck criterion was considered. In Fig. 1, failure criteria implemented in this paper are shown

☐ HASHIN 3D ☐ PUCK	TENSILE	COMPRESSION
FIBER	$F_{ft} = \left(\frac{\sigma_{11}}{X_t}\right)^2 + \alpha_f \left[\left(\frac{\sigma_{12}}{S_{12}}\right)^2 + \left(\frac{\sigma_{13}}{S_{13}}\right)^2\right] \geq 1$	$F_{fc} = \left(\frac{\sigma_{11}}{X_c}\right)^2 \geq 1$
MATRIX	$F_{mt} = \left(\frac{\sigma_{22}+\sigma_{33}}{Y_t}\right)^2 + \frac{1}{S_{23}^2}(\sigma_{23}^2 + \sigma_{22}\sigma_{33}) + \left(\frac{\sigma_{12}}{S_{12}}\right)^2 + \left(\frac{\sigma_{13}}{S_{13}}\right)^2 \geq 1$	$F_{mc} = \left(\frac{\sigma_{nt}}{S_{23}^A - \eta_{nt}\sigma_{nn}}\right)^2 + \left(\frac{\sigma_{nl}}{S_{12} - \eta_{nl}\sigma_{nn}}\right)^2 \geq 1$

Figure 1. Hashin 3D failure criteria for tensile/compression fiber and tensile matrix. Puck failure criterion for compression matrix.

where X_T, X_C, Y_T, Y_C are the fiber and matrix tensile/compression, respectively; all parameters in Puck's criterion ca be found in [9]. The equivalent strain measures for tensile/compression in the longitudinal and transverse directions are defined as

$$\delta_{eq}^{ft} = l_c \sqrt{\langle\varepsilon_{11}\rangle^2 + \varepsilon_{12}^2 + \varepsilon_{13}^2}, \qquad \delta_{eq}^{mt} = l_c \sqrt{\langle\varepsilon_{22}\rangle^2 + \langle\varepsilon_{33}\rangle^2 + \varepsilon_{12}^2 + \varepsilon_{23}^2 + \varepsilon_{13}^2} \qquad (1)$$

$$\delta_{eq}^{fc} = l_c \langle-\varepsilon_{11}\rangle, \qquad \delta_{eq}^{mc} = l_c \sqrt{\langle-\varepsilon_{22}\rangle^2 + \langle-\varepsilon_{33}\rangle^2 + \varepsilon_{12}^2 + \varepsilon_{23}^2 + \varepsilon_{13}^2} \qquad (2)$$

The corresponding equivalent stress measures are given by

$$\sigma_{eq}^{ft} = \frac{l_c(\langle\sigma_{11}\rangle\langle\varepsilon_{11}\rangle + \sigma_{12}\varepsilon_{12} + \sigma_{13}\varepsilon_{13})}{\delta_{eq}^{ft}}, \qquad \sigma_{eq}^{mt} = \frac{l_c(\langle\sigma_{22}\rangle\langle\varepsilon_{22}\rangle + \langle\sigma_{33}\rangle\langle\varepsilon_{33}\rangle + \sigma_{12}\varepsilon_{12} + \sigma_{23}\varepsilon_{23} + \sigma_{13}\varepsilon_{13})}{\delta_{eq}^{mt}} \qquad (3)$$

$$\sigma_{eq}^{fc} = \frac{l_c\langle-\sigma_{11}\rangle\langle-\varepsilon_{11}\rangle}{\delta_{eq}^{fc}}, \qquad \sigma_{eq}^{mc} = \frac{l_c(\langle-\sigma_{22}\rangle\langle-\varepsilon_{22}\rangle + \langle-\sigma_{33}\rangle\langle-\varepsilon_{33}\rangle + \sigma_{12}\varepsilon_{12} + \sigma_{23}\varepsilon_{23} + \sigma_{13}\varepsilon_{13})}{\delta_{eq}^{mc}} \qquad (4)$$

The damage d in the post-peak branch is defined as

$$d = \frac{\delta_{eq}^u(\delta_{eq} - \delta_{eq}^0)}{\delta_{eq}(\delta_{eq}^u - \delta_{eq}^0)} \qquad (5)$$

where $\delta_{eq} = l_c\varepsilon_{eq}$ is the equivalent displacement and $\delta_{eq}^u = \frac{2G_\alpha}{\sigma_{eq}}$ is the equivalent displacement when the damage d=1; l_c is the characteristic length and it corresponds to the cubic root of Gauss point volume. It is possible to compute the secant stiffness matrix in the damaged state as

$$C_d = \frac{1}{\Delta} \begin{bmatrix} C_{11} & C_{12} & C_{13} & 0 & 0 & 0 \\ C_{21} & C_{22} & C_{23} & 0 & 0 & 0 \\ C_{31} & C_{32} & C_{33} & 0 & 0 & 0 \\ 0 & 0 & 0 & C_{44} & 0 & 0 \\ 0 & 0 & 0 & 0 & C_{55} & 0 \\ 0 & 0 & 0 & 0 & 0 & C_{66} \end{bmatrix} \qquad (6)$$

More detailed description of secant stiffness matrix in the damage state can be found in [8].

2D Carrera Unified formulation

CUF allows writing the equations of any refined theory 1D, 2D, or 3D in terms of a few fundamental nuclei FNs. The three-dimensional displacement field $u(x, y, z)$ can be expressed as a product between a 2D in-plane shape function $N_i(x, y)$ of order p and 1D expansion function $F_\tau(z)$:

$$u(x, y, z) = N_i(x, y)F_\tau(z)u_{\tau i} \qquad \tau = 1, \dots, M \qquad i = 1, \dots, p + 1 \qquad (7)$$

More detailed of 2D plate modelling in CUF can be found in Ref. [10]. The linear strain-displacement relation is then expressed as $\varepsilon = Bu$, where B is the differential operator. The constitutive relation is given by $\sigma = C^{sec}$, where C^{sec} is obtained from damage model described in the previous section.

Numerical results

In this section, two cases of numerical assessments are shown. The material system used is IM7/8552 carbon fiber reinforced polymer (CFRP) with ply thickness of 0.125 mm, Table 1.

Table 1. Material properties of the IM7/8552 carbon fiber reinforced polymer.

E_1 [GPa]	E_2 [GPa]	E_3 [GPa]	G_{12} [GPa]	G_{13} [GPa]	G_{23} [GPa]	v_{12}	v_{13}	v_{23}
150.0	11.0	11.0	5.8	5.8	2.9	0.34	0.34	0.48
X_T [MP]	X_C [MP]	Y_T [MPa]	Y_C [MPa]	S_{12} [MPa]	G_1^T [kJ/m²]	G_2^T [kJ/m²]	G_1^C [kJ/m²]	G_2^C [kJ/m²]
2560.0	1690.0	73.0	250.0	90.0	120.0	2.6	80.0	4.2

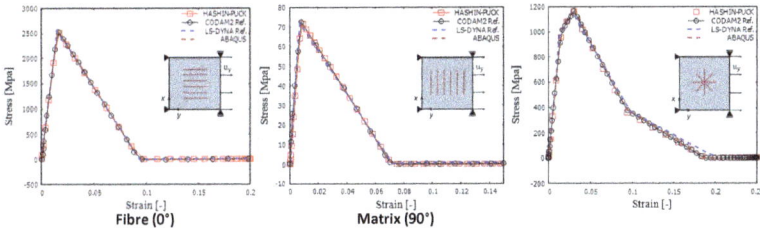

Figure 2. Stress-strain response of the single element under tensile loading with fiber orientation of 0°, 90° and [90/45/0/-45]₂ₛ, respectively. Comparison with ABAQUS response, CODAM2 Ref. [5] and LS-DYNA Ref. [11].

The first case refers to a single of size $1\ mm \times 1\ mm$ with under uni-axial strain conditions both for tensile and compression loading. Three tests are considered referring to loading in the longitudinal direction (along the fiber 0°), in the transverse direction (perpendicular to the fiber 90°) and considering a quasi-isotropic [90/45/0/-45]₂ₛ laminate. The in-plane discretization consists of one Q4 element and each ply thickness is modeled via a single linear (LE1) Lagrange polynomial. In Fig. 2, the stress-strain response of the single element under tensile loading with fiber orientation of 0°, 90° and [90/45/0/-45]₂ₛ laminate, respectively. The response is obtained by Hashin-Puck damage model correlates well with numerical reference CODAM2 [5], LS-DYNA [11] and ABAQUS model. Similarly, the stress-strain response of the single element under compressive loading is shown in Fig. 3, compared to ABAQUS, CODAM2 [8] and LS-DYNA [11] response.

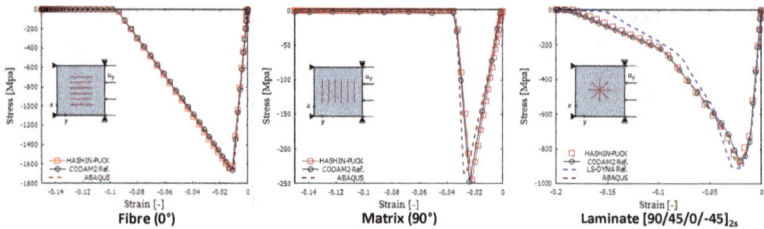

Figure 3. Stress-strain response of the single element under compression loading with fiber orientation of 0°, 90° and [90/45/0/-45]$_{2s}$, respectively. Comparison with ABAQUS response, CODAM2 Ref. and LS-DYNA Ref [5].

The next numerical assessment refers to an open-hole [45/90/-45/0]$_{2s}$ laminates of size $32\ mm \times 32\ mm$ subjected to compressive loads. In Fig. 4, the predictive failure strength computed with Hashin-Puck damage model and compared with ABAQUS, CODAM2 [8] and experiment [12] results.

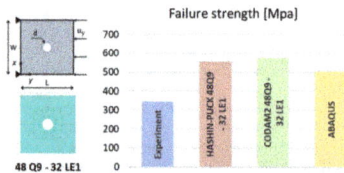

Figure 4. Comparison of failure strength of the open-hole [45/90/-45/0]$_{2s}$ specimen compared with experiment [12], CODAM2 [8] and ABAQUS results.

Conclusions

The current work presents the development of damage model based on Hashin 3D failure criteria and Puck's failure criteria for matrix compression using higher-order structural theories based on CUF. Three assessments are shown refers to a single elements, under tensile and compressive loading, and an open-hole specimen under compressive loading. The predictions of the proposed framework were in good general agreement with reference numerical and experimental results. The validation of this damage model highlights the advantages of CUF in computational costs. Future investigations include extending the present work to more complex structural problems. Other future works include modeling progressive failure of composite across scales through development of computationally-efficient advanced structural models, as micromechanical progressive failure analysis.

References

[1] I. Kaleel, M. Petrolo and E. Carrera, Elastoplastic and progressive failure analysis of fiber-reinforced composites via an efficient nonlinear microscale model, Aerotecnica Missili & Spazio, vol. 97, pp. 103-110, 2018.

[2] P. Ladeveze and E. LeDantec, Damage modelling of the elementary ply for laminated composites, Composites Science and Technology, vol. 43, no. 3, pp. 257-267, 1992. https://doi.org/10.1016/0266-3538(92)90097-M

[3] K. V. Williams, R. Vaziri and A. Poursartip, A physically based continuum damage mechanics model for thin laminated composite structures, International Journal of Solids

and Structures, vol. 40, no. 9, pp. 2267-2300, 2003. https://doi.org/10.1016/S0020-7683(03)00016-7

[4] A. Forghani, N. Zobeiry, A. Poursartip and R. Vaziri, A structural modelling framework for prediction of damage development and failure of composite laminates, Journal of Composite Materials, vol. 47, no. 20, pp. 2553-2573, 2013. https://doi.org/10.1177/0021998312474044

[5] M. H. Nagaraj, J. Reiner, R. Vaziri, E. Carrera and M. Petrolo, Progressive damage analysis of composite structures using higher-order layer-wise elements, Composites Part B: Engineering, vol. 190, 2020. https://doi.org/10.1016/j.compositesb.2020.107921

[6] L. Bek, R. Kottner and V. Laš, Material model for simulation of progressive damage of composite materials using 3D Puck failure criterion, Composite Structures, vol. 259, 2021. https://doi.org/10.1016/j.compstruct.2020.113435

[7] E. Carrera, M. Cinefra, M. Petrolo and E. Zappino, Finite element analysis of structures through unified formulation, John Wiley & Sons, 2014.

[8] M. H. Nagaraj, J. Reiner, R. Vaziri, E. Carrera and M. Petrolo, Compressive damage modeling of fiber-reinforced composite laminates using 2D higher-order layer-wise models, Composites Part B: Engineering, vol. 215, 2021. https://doi.org/10.1016/j.compositesb.2021.108753

[9] X. Li, D. Ma, H. Liu, W. Tan, X. Gong, C. Zhang and Y. Li, Assessment of failure criteria and damage evolution methods for composite laminates under low-velocity impact, Composite structures, vol. 207, pp. 727-739, 2019. https://doi.org/10.1016/j.compstruct.2018.09.093

[10] E. Zappino, G. Li, A. Pagani and E. Carrera, Global-local analysis of laminated plates by node-dependent kinematic finite elements with variable ESL/LW capabilities, Composite Structures,, vol. 172, pp. 1-14. https://doi.org/10.1016/j.compstruct.2017.03.057

[11] J. Reiner, T. Feser, D. Schueler, M. Waimer and R. Vaziri, Comparison of two progressive damage models for studying the notched behavior of composite laminates under tension, Composite Structures, vol. 207, pp. 385-396, 2019. https://doi.org/10.1016/j.compstruct.2018.09.033

[12] J. Lee and C. Soutis, Measuring the notched compressive strength of composite laminates: Specimen size effects, Composites Science and Technology, vol. 68, no. 12, pp. 2359-2366, 2008. https://doi.org/10.1016/j.compscitech.2007.09.003

Aerospace Science and Engineering - IV Aerospace PhD-Days   Materials Research Forum LLC
Materials Research Proceedings 42 (2024) 66-71   https://doi.org/10.21741/9781644903193-15

Simulation of post-grasping operations in closed-chain configuration using Kane's method

David Paolo Madonna[1,a] *

[1]Department of Mechanical and Aerospace Engineering, Sapienza University of Rome, Via Eudossiana 18, 00184 Rome, Italy

[a]davidpaolo.madonna@uniroma1.it

Keywords: Post-Grasping Operations, Multibody Closed-Chain Configuration, Space Manipulator System, Kane's Method

Abstract. This work focuses on techniques that enable the modeling of a multibody spacecraft in a closed-chain configuration. This scenario pertains to a space manipulator system equipped with two or more robotic arms that have grasped a target object. The dynamic equations of the multibody system, consisting of the chaser and target satellites, are derived using a Kane's formulation for nonholonomic constrained systems. This formulation eliminates the need for including Lagrange's multipliers in the set of equations. Numerical simulations of a post-grasping maneuver between a space manipulator system and a target satellite are conducted to validate the proposed formulation.

Introduction

Modeling multibody spacecraft in closed-chain configurations implies a higher level of complexity with respect to tree configurations. Various approaches are reviewed in Ref. [1], where an extensive comparison is conducted. Among all the strategies presented in this reference, the "cut joint" approach stands out as a compromise between physical consistency, measured in terms of conservation of mechanical energy (in the absence of dissipative effects), and computational efficiency. In this approach, a closed-chain is severed at an underactuated joint to form two separate branches in an open-chain configuration. At the cut joint, a holonomic constraint must be enforced to ensure the compatibility. Different strategies exist to derive the dynamical equations for such systems. When employing a Lagrange's approach, this results in a system of differential-algebraic equations (DAE), where the Lagrange multipliers serve as the adjoint unknowns to recover the internal actions provided by the kinematic constraints. However, a viable alternative is suggested in Ref. [2], introducing a novel version of Kane's equations tailored for constrained systems. This yields a dynamic system divided into two distinct components: an Ordinary Differential Equation (ODE) system comprising motion equations, and an algebraic system of constraint equations. The computation of the latter is only undertaken when explicitly required by the analysis tasks. In the context of this study, this Kane's formulation is employed to simulate a post-grasping scenario. The preparation for repairing a spinning out-of-service satellite is examined, considering a dual-arm space manipulator system (SMS) [3-4]. Initially, the de-spinning process is executed, followed by reducing the relative distance between the target and chaser to a value that permits repairing.

Kane's formulation for constrained systems

The Kane's formulation reported in reference [2] can be applied to model mechanical systems characterized both by kinematic and nonholonomic constraints. However, kinematic constraints must first undergo a time derivation to be transformed into velocity constraints. A multibody system characterized by n generalized coordinated q_i (collected in the column vector $\underline{q}$) and n generalized velocities u_i (collected in the column vector $\underline{u}$), which are linearly dependent on the

Aerospace Science and Engineering - III Aerospace PhD-Days   Materials Research Forum LLC
Materials Research Proceedings 42 (2024) 66-71   https://doi.org/10.21741/9781644903193-15

time derivative of the generalized coordinates [5], is considered. Then, if *(n-p)* velocity constraints are imposed, the set of n generalized velocities can be divided into two subsets: $\underline{u_I} = [u_1 \cdots u_p]^T$ and $\underline{u_D} = [u_{p+1} \cdots u_n]^T$ containing the independent and the dependent generalized velocities respectively. The constraint equations can be expressed as

$$\underline{u_D} = A\left(\underline{q}, t\right)\underline{u_I} + \underline{b}\left(\underline{q}, t\right),\tag{1}$$

where $A \in \mathbb{R}^{(n-p) \times p}$ and $\underline{b} \in \mathbb{R}^{n-p}$ both depend on the generalized coordinates and time t. It can be proven that, under the constraint equations (1), the Kane's dynamical equations can be written as

$$\tilde{\underline{F}} + \tilde{\underline{F}}^* = A_2(\underline{F} + \underline{F}^*) = \underline{0}\tag{2}$$

where $\tilde{\underline{F}}, \tilde{\underline{F}}^* \in \mathbb{R}^p$ are the vectors of generalized active forces and generalized inertia forces [2] respectively defined for the constrained system, while $\underline{F}, \underline{F}^* \in \mathbb{R}^n$ are referred to the unconstrained system, and $A_2 = [I_{p \times p} \quad A^T]$ where $I_{p \times p}$ is the identity matrix of dimension p. The generalized inertia force vector can be rewritten as

$$\underline{F}^* = -M(\underline{q}, t)\underline{\dot{u}} - \underline{nl}(\underline{q}, \underline{u}, t)\tag{3}$$

where $M \in \mathbb{R}^{n \times n}$ is the generalized mass matrix, while $\underline{nl} \in \mathbb{R}^n$ is the vector of nonlinear terms of dynamics, i.e. the terms that do not linearly depend on the time derivative of the generalized velocities. Substituting Eq. (3) in Eq. (2) one obtains

$$A_2 M \underline{\dot{u}} = -A_2 \underline{nl} + A_2 \underline{F}.\tag{4}$$

To incorporate constraint equation (1) into the system, it needs to undergo a time derivation to be expressed in the form of acceleration:

$$\underline{\dot{u}_D} = A\underline{\dot{u}_I} + \dot{A}\underline{u_I} + \underline{\dot{b}} \quad \Rightarrow \quad A_1\underline{\dot{u}} = \dot{A}\underline{u_I} + \underline{\dot{b}}\tag{5}$$

where $A_1 = [-A \quad I_{(n-p) \times (n-p)}]$. Hence, merging of Eqs. (4) and (5) leads to

$$T\underline{\dot{u}} = \left(\left[\dot{A}\underline{u_I} + \underline{\dot{b}}\right]^T \quad \left[A_2\left(-\underline{nl} + \underline{F}\right)\right]^T\right)^T\tag{6}$$

where the matrix $T = \left[A_1{}^T \quad (A_2 M)^T\right]^T \in \mathbb{R}^{n \times n}$ is invertible (the proof is provided in Ref. [2]). Eq. (6) fully describes the motion of a constrained system, yet he does not offer any information about the constraint reactions. However, they can be easily evaluated after solving Eq. (6) through a back-substitution procedure.

Application to the close-chain multibody configuration resolved via cut joint approach
In the framework of Kane's formulation, the dynamical equations for a closed-loop multibody spacecraft can be obtained through the cut joint approach [1]. The process involves opening a closed chain into two open chains at the location of an underactuated joint, while simultaneously enforcing kinematic constraints that are equivalent to the removed joint. The derivation of the dynamical equations is now presented for the specific case of a slider-crank mechanism However, it's worth noting that this approach is applicable to any closed-chain structure.

(a)   (b)

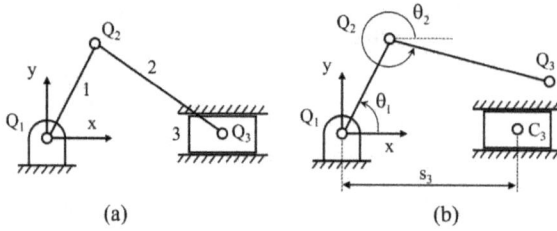

Figure 1: sketch of a slider-crank mechanism (a) before and (b) after the cut open procedure

The sketch of a slider-crank mechanism is depicted in Fig. 1, (a) before and (b) after the cut-open procedure. In the open configuration, the generalized coordinate vector is $q = [\theta_1 \quad \theta_2 \quad s_3]^T$, where the angles θ_1, θ_2 and the displacement s_3 are depicted in Fig. 1.b, and the generalized velocities are chosen to be the time derivative of the generalized coordinates. Following the Lagrange's or the standard Kane's methodology to derive the system dynamical equations [6], one obtains the same result, i.e.

$$M\dot{u} + \underline{nl} = \underline{F} + \underline{F}_{cr} \tag{7}$$

where $\underline{F}_{cr}$ is the generalized constraint reactions vector associated to the kinematic constraints that must be imposed to guarantee that the position and the velocity of Q_3 always coincide position and velocity of C_3. These constraints can be written in the following form:

$$\underline{\Phi}(\underline{q}) = \underline{0} \quad \Rightarrow \quad \begin{cases} l_1 \cos\theta_1 + l_2 \cos\theta_2 - s_3 = 0 \\ l_1 \sin\theta_1 + l_2 \sin\theta_2 = 0 \end{cases} \tag{8}$$

where l_1 and l_2 are the lengths of the two links. Now, Eq. (8) can be incorporated in Eq. (7) using Lagrange's multipliers or following the Kane's logic described in the previous Section. In the first case, Eq. (8) is derived twice with respect to time to obtain

$$\ddot{\underline{\Phi}} \equiv D\dot{\underline{u}} = \underline{\gamma} \tag{9}$$

where $D \in \mathbb{R}^{(n-p) \times n}$ and $\gamma \in \mathbb{R}^{n-p}$. For this case study $p=1$ and $n=3$. Since the Lagrange's multipliers vector $\underline{\lambda} \in \mathbb{R}^{n-p}$ is defined such that

$$\underline{F}_{cr} = -D^T \underline{\lambda} , \tag{10}$$

one finally obtains the following DAE system:

$$\begin{bmatrix} M & D^T \\ D & 0 \end{bmatrix} \begin{Bmatrix} \dot{u} \\ \underline{\lambda} \end{Bmatrix} + \begin{Bmatrix} \underline{nl} \\ \underline{0} \end{Bmatrix} = \begin{Bmatrix} \underline{F} \\ \underline{\gamma} \end{Bmatrix} , \tag{11}$$

This system is composed of five differential-algebraic equations. On the other hand, when utilizing the presented Kane's method, initially, only one time derivative of Eq. (8) is performed to obtain

$$\dot{\underline{\Phi}} \equiv C\underline{u} = \underline{\beta}, \tag{12}$$

where $C \in \mathbb{R}^{(n-p) \times n}$ and $\beta \in \mathbb{R}^{n-p}$. It is worth noting that Eq. (12) can be reconducted to the form of Eq. (1). In fact, for the case of kinematic constraints imposed in the cut joint approach, both $\underline{b}$ and $\underline{\beta}$ are null vectors, so

$$C\underline{u} = \begin{bmatrix} C_I & C_D \end{bmatrix} \begin{bmatrix} \underline{u_I} \\ \underline{u_D} \end{bmatrix} = C_I \underline{u_I} + C_D \underline{u_D} = \underline{0} \quad \Rightarrow \quad A = -C_D^{-1} C_I \tag{13}$$

where $C_D \in \mathbb{R}^{(n-p) \times (n-p)}$ and $C_I \in \mathbb{R}^{(n-p) \times p}$. Hence, following the steps described in the previous Section, one obtains a system which expression coincides with Eq. (6). Differently from Eq. (11), the obtained system is composed of three ordinary differential equations that completely describe the motion of the mechanism.

Numerical simulation of post-grasping scenario

In this section, numerical results for a post-grasping scenario are presented. The simulation initiates after the dual-arm Space Manipulator System (SMS) has successfully grasped a target with a mass of 100 kg and a radius of 1 m, which is rotating 5 °/s. Due to the rotational motion of the target, the grasping maneuver was executed after the chaser had synchronized its rotational motion with that of the target. The illustration of the SMS and its physical properties is provided in Fig. 2 and Tables 1-2, respectively. The sequence of operations follows the subsequent steps: first, the chaser performs a de-spinning maneuver while maintaining the target at 1.5 m. This phase lasts for 180 seconds. Then, after 60 additional seconds where the final state of the de-spinning phase is maintained, a manipulation maneuver is carried out. During this maneuver, that lasts 120 seconds, the relative distance between the target and the chaser is reduced from 1.5 m to 0.75 m. The mission concludes after another 240 seconds, during which the spacecraft maintains the desired final state. Numerical results are reported in Figs. 3-5. The results show that both the de-spinning and the manipulation of the target are successfully achieved within the depicted time span. Furthermore, the control efforts are modest for this mission, facilitating straightforward implementation in terms of actuator sizing.

Concluding remarks

The problem of modeling closed-chain multibody spacecraft has been addressed. The proposed approach relies on (i) a formulation of Kane's equations for constrained systems and (ii) the cut joint procedure. The former enables a reduction in the dimensionality of the equation system while retaining an Ordinary Differential Equation (ODE) structure. Meanwhile, the latter represents a compromise between numerical efficiency and the physical coherence of the results. The process of applying the cut joint technique within the Kane's formulation for constrained systems framework has been outlined. A numerical simulation has been conducted to showcase the capabilities of the presented approach.

Table 1: BUS specifications	
Mass [kg]	1000
J_1 [kg m^2]	590.83
J_2 [kg m^2]	270.83
J_3 [kg m^2]	486.67

Table 2: links specifications	
Length [m]	1
Radius [cm]	3.5
Mass [kg]	30

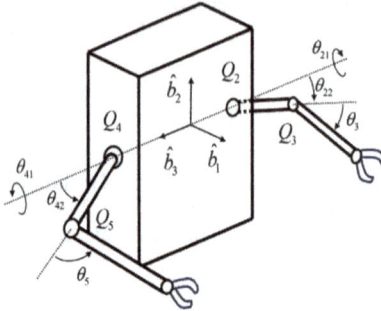

Figure 2: sketch of the Space
Manipulator System

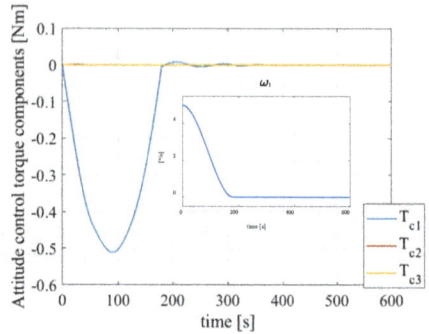

Figure 3: Attitude control torque components
and absolute angular rate along b_1

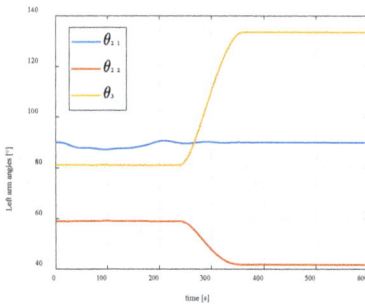

Figure 4: angles of the left arm

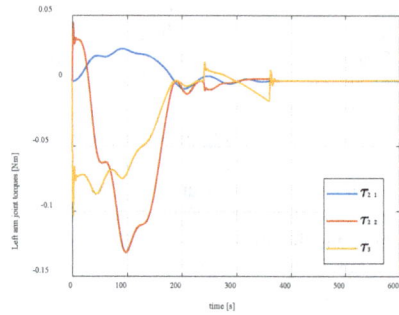

Figure 5: Attitude control torque components

References

[1] F. Marques, I. Roupa, M. Silva, P. Flores, H. Lankarani, Examination and comparison of different methods to model closed loop kinematic chains using Lagrangian formulation with cut joint, clearance joint constraint and elastic joint approaches, Mechanism and Machine Theory, Vol. 160, Elsevier, 2021. https://doi.org/10.1016/j.mechmachtheory.2021.104294

[2] A. Bajodah, D. Hodges, Ye-Hwa Chen, New Form of Kane's Equations of Motion for Constrained Systems, Journal of Guidance, Control, and Dynamics, Vol. 26, Is. 1, Pages 79-88, AIAA, 2003. https://doi.org/10.2514/2.5017

[3] A. De Bonis, F. Angeletti, P. Iannelli, P. Gasbarri, Mixed Kane–Newton Multi-Body Analysis of a Dual-Arm Robotic System for On-Orbit Servicing Missions, Aerotecnica Missili & Spazio, Vol. 100, Pages. 375–386, 2021. https://doi.org/10.1007/s42496-021-00088-0

[4] A. Stolfi, P. Gasbarri, M. Sabatini, Performance Analysis and Gains Tuning Procedure for a Controlled Space Manipulator Used for Non-cooperative Target Capture Operations, Aerotecnica Missili & Spazio, Vol. 97, Pages. 3-12, 2018. https://doi.org/10.1007/BF03404759

[5] T. R. Kane, D. A. Levinson, Dynamics: Theory and Applications, McGraw-Hill, 1985.

[6] A. Pisculli, L. Felicetti, M. Sabatini, P. Gasbarri, G. B. Palmerini, A Hybrid Formulation for Modelling Multibody Spacecraft, Aerotecnica Missili & Spazio, Vol. 94, Pages. 91–101, 2015. https://doi.org/10.1007/BF03404692

Aerospace Science and Engineering - IV Aerospace PhD-Days   Materials Research Forum LLC
Materials Research Proceedings 42 (2024) 72-75   https://doi.org/10.21741/9781644903193-16

Nonlinear dynamic modelling and performance of deployable telescopic tubular mast (TTM)

Tongtong Sun[1,a], Paolo Gasbarri[2,b], Lin Du[1,c], Zichen Deng[3,d]

[1]School of Mathematics and Statistics, Northwestern Polytechnical University, Xi'an 710072, China

[2]School of Aerospace Engineering, Sapienza University of Rome, Rome, 00138, Italy

[3]School of Mechanics, Civil Engineering and Architecture, Northwestern Polytechnical University, Xi'an 710072, China

[a]tongtongsun@mail.nwpu.edu.cn, [b]paolo.gasbarri@uniroma1.it, [c]lindu@nwpu.edu.cn, [d]dweifan@nwpu.edu.cn

Keywords: Flexible Vibration, Nonlinear Dynamics, Telescopic Tubular Mast, Deploying Process

Abstract. The aim of this work is to present the longitude and transverse vibrations of a deployable TTM, which is attached on a spacecraft system, considering the effect of rigid-flexible coupling phenomenon. The proposed model is derived based on the principle of virtual work and discretized by assumed mode method. To introduce the nonlinear effect, the von Kármán strain is adopted. Additionally, locking and restart behaviors are taken into account in the modelling procedure of the deploying process. Finally, the dynamic phenomena of the longitude and transverse displacements are analyzed at different deploying velocities.

Introduction

As one of the effective ways to achieve large constructions in space, the deployable Telescopic Tubular Mast (TTM) is developed by Northrop Grumman and widely used in the spacecraft systems [1]. The TTM is designed to have several tubes with different radius and each one is nested within one another at the launch stage. When deployed, the tube sections will unfold step by step and locked after each section's motion, experience orbital and attitude rotation at the same time, which results in a hard behavior to model. To solve this problem, a ground test was conduced to determine the deployed stiffness, natural frequencies and its load capability [2]. Moreover, an axially moving cantilever beam model was used when designing the control system to achieve the deployment criteria. However, only the transverse deformation was considered in their work. The longitude motion is also non-neglected when the deployable structures undergoing rotation motions [3]. Therefore, the longitude and transverse vibrations of the TTM during the deployment process are investigated in this work. Moreover, the stepped characteristics of the TTM is considered in this work instead of the uniform beam assumption to make the model more accurate.

Modeling process

The TTM considered in this work is composed of 17 flexible tube sections with different radius and 1 rigid container. All deployable tubes are nested in the container at the initial state and extended sequentially. The spacecraft system contains one rigid central cube hub and two deployable TTMs. Considering the deployable TTM in space as a stepped Euler-Bernoulli beam with axially moving motion (Fig. 1). The length of the deployed structure is $l(t)$, which is varying with time. The other symbols and corresponding values used in this work are listed in Table 1.

Aerospace Science and Engineering - III Aerospace PhD-Days
Materials Research Proceedings 42 (2024) 72-75

Materials Research Forum LLC
https://doi.org/10.21741/9781644903193-16

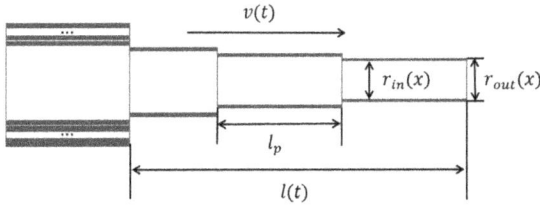

Fig. 1 Deploying process of the TTM

Table 1. Symbols descriptions

Symbols	Description	Value
l_p [m]	length of each section	3
E [GPa]	Young's modulus	70
$I(x)$ [kg·m^2]	moment of inertia	\
ρ [kg/m^3]	density of the boom	2700
$A(x)$ [m^2]	cross section of the boom	\
a [m]	half-length of the central hub	2
r [km]	orbital radius	6700

To describe the nonlinear coupling dynamics of the deployable boom, the von Kármán strain is adopted in this formulation. Therefore, the nonlinear normal strain is expressed as:

$$\varepsilon_{xx} = \frac{\partial u(x,t)}{\partial x} - y\frac{\partial^2 w(x,t)}{\partial x^2} + \frac{1}{2}(\frac{\partial w(x,t)}{\partial x})^2 \tag{1}$$

where $u(x,t)$ is the longitude displacement in the x directions, and $w(x,t)$ is the transverse displacement in the y direction. Due to the fact that the two deployable TTMs will not affect each other on flexible vibration [4], only one side is considered in this work. The nonlinear coupling dynamic equations are obtained based on the principle of virtual work:

$$\frac{D^2 u}{Dt^2} + \dot{v} - w\ddot{\alpha} - 2(\dot{\alpha}+\dot{\theta})\frac{Dw}{Dt} - (a+l_p+x+u)(\dot{\alpha}+\dot{\theta})^2 - \frac{EA}{\rho A}\frac{\partial^2 u}{\partial x^2}$$
$$= \frac{\mu}{2r^3}(a+l_p+x)(1+3\cos 2\alpha) \tag{2}$$

$$\frac{D^2 w}{Dt^2} + (a+l_p+x+u)\ddot{\alpha} + 2(v+\frac{Du}{Dt})(\dot{\alpha}+\dot{\theta}) - (\dot{\alpha}+\dot{\theta})^2 w - (P_x\frac{\partial w}{\partial x} + P\frac{\partial^2 w}{\partial x^2})$$
$$+ \frac{EI}{\rho A}\frac{\partial^4 w}{\partial x^4} - \frac{EA}{\rho A}(\frac{\partial^2 u}{\partial x^2}\frac{\partial w}{\partial x} + \frac{\partial^2 w}{\partial x^2}\frac{\partial u}{\partial x}) = -\frac{3\mu}{2r^3}(a+l_p+x)\sin 2\alpha \tag{3}$$

where α is the attitude angle of the spacecraft system, and θ is the orbital angle. The right sides of Eqs. (2) and (3) represent the perturbation of gravity gradient associated with the rotational motion of the spacecraft system. Using the assumed mode method, the functions for longitude and transverse displacement can be expressed as the summation of mode functions and their corresponding amplitudes:

$$\begin{cases} u(x,t) = \sum_{i=1}^{N_1} \psi_i(x)q_i(t) \\ w(x,t) = \dfrac{1}{\sqrt{l(t)}} \sum_{j=1}^{N_2} \varphi_j(x)p_j(t) \end{cases}$$

$$(4)$$

where N_1, N_2 are the truncated order according to the Galerkin's method. The mode shape function can be written as follows according to the boundary conditions [3]:

$$\begin{cases} \psi_i(\varepsilon) = \sin(\dfrac{2i-1}{2}\pi\varepsilon) \\ \varphi_j(\varepsilon) = \cosh \lambda_j\varepsilon - \cos \lambda_j\varepsilon - B(\sinh \lambda_j\varepsilon - \sin \lambda_j\varepsilon) \end{cases}$$

$$(5)$$

where $\varepsilon = \dfrac{x}{l(t)}$, $B = \dfrac{\cos \lambda_i + \cosh \lambda_i}{\sin \lambda_i + \sinh \lambda_i}$, λ_i is the ith solution of $1 + \cos \lambda_i \cosh \lambda_i = 0$.

Results and discussions

To analyze the effects of the deploying velocities on longitude and transverse displacements, two constant values denoted as $v = 0.1\,m/s$ and $v = 0.5\,m/s$ are adopted in this section when the pause time is $0.05s$, and the spacecraft system remains sun-facing. Here, the pause time represents the time gap after each section is expanded. As a results, the functions of length, velocity and acceleration of the boom are no longer continuous functions due to the existence of time interval. The variations in velocity and acceleration are shown in Fig 2, with the velocity set at 0.5m/s and time pause time at 2s the discontinuous characteristics of the deployment process clearly visible.

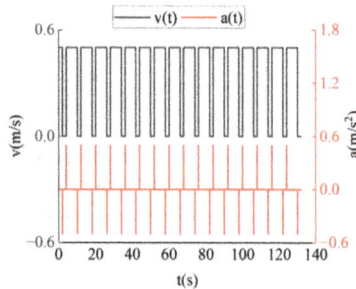

Fig. 2 Velocity and acceleration of the deploying process

The longitudinal and transverse displacements of the tip of the boom under various deploying velocities are depicted in Fig. 3. Throughout the deployment process, the longitudinal displacement remains very small (order of 10^{-3} mm). As the velocity increases, the longitudinal displacement also increases. The intermittent restart and stop actions during the unfolding process of each section lead to sudden changes in amplitude, attributed to the additional pulse acceleration applied along the x-axis. The impact of velocity on transverse vibration is more pronounced, with displacement decreasing as velocity increases. This is because higher velocities correspond to greater transient stiffness of the boom. In addition, the displacement in y-direction is induced by the gravity gradient force. Under the same structural length, the higher deploying velocity induce the lower gravity gradient force since the larger the attitude angle corresponding to the larger sin function when the attitude angle is less than $\pi / 4$ rad.

(a) $v = 0.1 \, m/s$

(b) $v = 0.5 \, m/s$

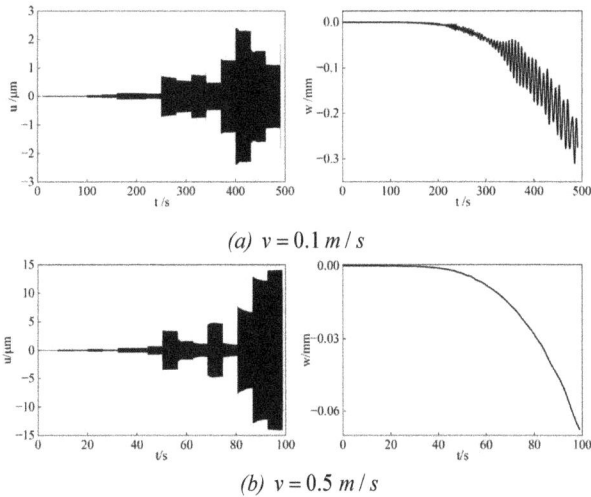

Fig 3. Longitude and transverse displacements during the deploying process for different deploying velocities

Conclusions

The numerical simulation on the unfolding process of TTM is conducted in this work considering the coupling effects of attitude motion. The longitude and transverse displacements are investigated under different deploying velocity. Numerical simulations demonstrate that the longitude displacement increased with the increase of the velocity. However, the transverse displacement exhibit opposite behavior due the properties of dynamical natural frequency and the effects of the gravity gradient force.

References

[1]. X. Zhang, R. Nie, Y. Chen, et al., Deployable structures: structural design and static/dynamic analysis, J. Elasticity. 146 (2021) 1-37. https://doi.org/10.1007/s10659-021-09860-6.

[2]. M. Mobrem and C. Spier. Design and Performance of the Telescopic Tubular Mast. The 41st Aerospace Mechanisms Symposium (2012) 127-140.

[3]. S. Park, H.H. Yoo and J. Chung, Vibrations of an axially moving beam with deployment or retraction, AIAA J. 51 (2013) 686-696. https://doi.org/10.2514/1.J052059

[4]. T.T. Sun, S. Zhang, L. Du, et al., Nonlinear dynamic modeling for analysis of large spacecraft with extendible appendages, Appl. Math. Model. 123 (2023) 526-545. https://doi.org/10.1016/j.apm.2023.06.041

Aerospace Science and Engineering - IV Aerospace PhD-Days   Materials Research Forum LLC
Materials Research Proceedings 42 (2024) 76-79   https://doi.org/10.21741/9781644903193-17

A short overview of hydrogen storage for sustainable aviation

Sergio Bagarello[1,2,a], Dario Campagna[1,b], Alberto Milazzo[1,2,c], Ivano Benedetti[1,2,d*]

[1]Department of Engineering, Università degli Studi di Palermo, Viale delle Scienze, Edificio 8, 90128 Palermo, Italy

[2]Sustainable Mobility Center - Centro Nazionale per la Mobilità Sostenibile – MOST, Italy

[a]sergio.bagarello@unipa.it, [b]dario.campagna@unipa.it, [c]alberto.milazzo@unipa.it, [d]ivano.benedetti@unipa.it

Keywords: Aviation Hydrogen, Aircraft Hydrogen Storage, High-Pressure Vessels, Cryogenic Tanks, Sustainable Aviation

Abstract. The aviation industry is facing environmental challenges, prompting the need for aircraft technologies aligned with climate goals. Hydrogen has emerged as a clean fuel option, offering significant advantages in supply potential and sustainability. Reliable and safe storage technologies are crucial for integrating this fuel into aircraft design. This overview examines current storage technologies for hydrogen in aviation, emphasizing tank materials and structural aspects to enable sustainable fuel use. Requirements for integrating fuel systems into existing or new aircraft architectures are discussed, considering both gaseous and liquid hydrogen storage. Both high pressures and cryogenic storage conditions are explored. Materials for hydrogen tanks are reviewed, with a focus on improving efficiency and addressing issues like thermal insulation and hydrogen embrittlement.

Introduction

Surrounded by growing concerns over global warming, the aviation industry is under pressure to adopt sustainable and zero-emission propulsion technologies. Biofuels like Sustainable Aviation Fuel (SAF) offer promise but face challenges in cost and availability. Electric and hybrid-electric aviation show potential benefits, yet battery limitations remain as a crucial bottleneck [1]. Hydrogen emerges as a highly appealing option due to its availability and sustainability, even if strongly affected by production methods. Addressing aviation's environmental impact requires a multifaceted approach, including technological advancements and infrastructure upgrades.

Hydrogen in aviation

Hydrogen has a long history in aviation, from early hot-air balloons to airships like the Zeppelins. The U.S. and Soviet Union made notable advancements in the past century, with the U.S. achieving the first flight powered by liquid hydrogen. Challenges such as logistics and fuel system integration persisted and in the 2000s, the EU funded the Cryoplane Project, demonstrating the feasibility of liquid hydrogen aircraft despite the need for better fuel tanks. Economic analyses suggest that while costs may initially rise, liquid hydrogen could become the most cost-effective renewable aviation fuel by 2040 [2].

Onboard hydrogen storage

Hydrogen provides three times the specific energy of traditional jet fuel (120 MJ/kg vs 43.2 MJ/kg), emitting no CO_2 when combusted, making it suitable to meet environmental goals. Despite its flammability, it ignites faster and offers superior cooling properties due to its high thermal conductivity, heat capacity, and low viscosity. It can be produced on-site with zero emissions through various thermochemical processes [3]. However, its low energy density per volume (10.1 MJ/l) requires about four times the storage volume of traditional fuel and often demands cryogenic temperatures to address spatial integration challenges.

Aerospace Science and Engineering - III Aerospace PhD-Days   Materials Research Forum LLC
Materials Research Proceedings 42 (2024) 76-79   https://doi.org/10.21741/9781644903193-17

Aircraft layout considerations

Adapting aircraft for hydrogen usage is predicted to involve significant modifications, particularly in storage systems. Hydrogen's low volumetric energy density complicates storage, as conventional wing-integrated tanks are impractical. Alternative configurations, like fuselage placed tanks, are currently being explored. Future perspectives include blended-wing-body designs for both structural and aerodynamic efficiency, but their adoption requires an industry-wide shift in paradigm in design, maintenance, and operation. The choice between integral and non-integral tanks, Fig (1), is crucial: the first category offers efficiency and easier inspections, while the latter allows for modular designs.

Figure 1 - Integral vs. Non-integral hydrogen tank

Hydrogen storage technologies

Hydrogen can be stored in gaseous (GH_2), liquid (LH_2) or subcooled liquid (sLH_2) forms or through material-based approaches. Compared to traditional fuels, hydrogen storage poses challenges due to its low volumetric energy density. Hydrogen tanks require careful design encompassing geometric, mechanical, and thermal constraints, as well as the consideration of the aircraft's specific mission profile [4]. The tank *gravimetric efficiency* defined as

$$\eta_{tank} = \frac{W_{H_2}}{W_{H_2} + W_{tank}} \tag{1}$$

is a crucial metric for investigating the efficiency of the considered storage technology for aviation. In Eq.(1), W_{H_2} represents the hydrogen (fuel) weight, while W_{tank} represents the tank structural weight. Higher η_{tank} values are preferred, indicating a lower weight penalty for the hydrogen storage system. This parameter is crucial for assessing progress towards developing tanks comparable to conventional kerosene tanks, which typically have efficiencies approaching 100%, when integrated tanks are employed. Poor gravimetric efficiencies result in increased operating empty weight and performance penalties, leading to higher overall mission energy consumption. The gravimetric efficiency can be enhanced increasing the hydrogen volumetric density either through high pressure (GH_2) or low temperature (LH_2).

Gaseous hydrogen storage

Hydrogen's low volumetric energy density requires significant compression for aircraft use, typically at pressures ranging from 350 to 700 bar. Tanks must balance the increase of energy density through high pressure with structural efficiency. In this sense carbon-fiber-reinforced-plastic (CFRP) or glass-reinforced-plastic (GRP) composites offer lightweight options. Various tank geometries distribute pressure loads, reducing weight penalties. Gaseous hydrogen tanks range from metal to liner-less composites, each with different efficiencies. High-pressure vessels use composites or metals with liners to prevent gas leakage. Type-V tanks without liners require careful design and testing to avoid microcracks and leaks [5]. Exploration of thermoplastic matrices in composites offers recycling advantages for hydrogen fuel tanks, further enhancing sustainability.

Liquid hydrogen storage

Storing hydrogen at cryogenic temperatures increases fuel density, allowing for lower storage pressures (~2 bar). Liquid hydrogen offers the highest volumetric storage density (~50%) compared to other methods [5]. However, managing extreme temperature differentials and

Aerospace Science and Engineering - III Aerospace PhD-Days   Materials Research Forum LLC
Materials Research Proceedings 42 (2024) 76-79   https://doi.org/10.21741/9781644903193-17

preventing high boil-off rates are critical challenges. Slush hydrogen and subcooled LH_2 further increase density (~80 kg/m³). Effective thermal insulation minimizes fuel wastage due to venting. While compressed hydrogen tanks offer flexibility, liquid hydrogen tanks are found to be more efficient for larger quantities of hydrogen [6]. Integration of cryogenic tanks with aircraft systems allows for fuel management and component cooling simultaneously. Lightweight, well-insulated tanks are essential for widespread use of liquid hydrogen in aviation. Thermal insulation used in such components must minimize heat transfer while balancing mass and cost. Insulation methods include multilayer, vacuum jacket, and foam insulation, with external insulation often preferred to cope with H_2 corrosivity. Insulation thickness optimization is of concern, considering heat leak reduction and pressurization impacts as well as overall tank mass. The *multilayer insulation* (MLI) system consists of reflective foils and metal shields, alternating with insulating materials, to minimize heat transfer. *Vacuum jackets,* enhance insulation efficiency, but venting subsystems are necessary to seal the vacuumed region adding mass and complexity to the aircraft. Handling dimensional variations from thermal cycles during tank operations is a challenge due to thermal coefficient mismatches. *Foam insulation*, Fig.(2), though generally thicker than MLI, is found to be effective and can efficiently and passively insulate even with a single tank wall.

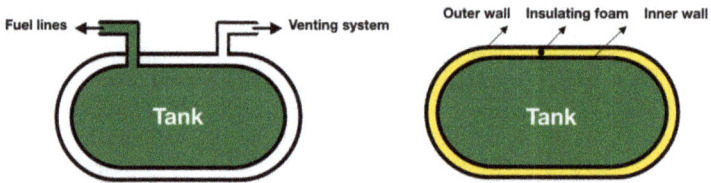

Figure 2 - Vacuum insulation (left) vs. Foam insulation (right)

Advanced materials, like *aerogels*, show promise for future applications, showing exceptional insulating properties. Implementing aerogel technology could lead to reduced weight, simplified design, enhanced safety, cost savings, and improved sustainability. However, specific tank design considerations and safety evaluations are necessary due to aerogel's unique characteristics. Ongoing research aims at establishing industry standards for competitive gravimetric efficiencies, safety, and low cost in liquid hydrogen tanks.

Hydrogen embrittlement

Hydrogen embrittlement is an example of environmentally assisted cracking caused by the very small size of H_2 molecules that weaken materials like steel and titanium in storage tanks subjected to internal pressure coupled with hydrogen exposure, leading to sudden failures even below the yield strength of a material [7]. Preventive measures are known and include the use of aluminum alloys under the form of liners, but understanding the interaction of liner-composite-insulation systems under cryogenic temperatures is important for large-scale introduction of hydrogen in aviation.

Improving tanks efficiency

Improving hydrogen tank efficiency leverages on using lightweight composite materials like carbon fiber and exploring advanced materials such as graphene and metal-organic frameworks for increasing storage performances. Structural health monitoring and strategically reduced safety factors can further boost tank efficiency. *Graphene* and graphene-like materials (GLMs) show promise for high-density hydrogen storage by enhancing storage capacity and kinetics. *Carbon nanotubes* (CNTs) offer potential for hydrogen adsorption at room temperature, with surface modification techniques aiming to improve absorption. *Metal-organic frameworks* (MOFs) offer high-density storage with rapid kinetics and controlled release properties. *Hollow glass*

Aerospace Science and Engineering - III Aerospace PhD-Days   Materials Research Forum LLC
Materials Research Proceedings 42 (2024) 76-79   https://doi.org/10.21741/9781644903193-17

microspheres (HGM) reduce the weight of hydrogen tanks and reinforce them while acting as insulating additives. However, their use requires specific heating conditions for hydrogen release, leading to increased energy consumption and costs. *Structural health monitoring* (SHM) systems and *non-destructive inspection* (NDI) methods can enhance safety and efficiency. Lead Zirconated Titanate (PZT) sensors show promise for withstanding cryogenic temperatures, while optical fibers offer real-time monitoring capabilities. Addressing these challenges is crucial for advancing efficient and safe hydrogen storage systems in aviation.

Conclusions

Hydrogen-powered aviation offers a promising path towards balancing economic growth and environmental responsibility. However, onboard hydrogen storage remains a significant challenge. Efforts are underway to enhance the performance of pressure vessels through innovative structural and insulation technologies. Liquid hydrogen appears as a promising option for large-scale commercial operations, although further research is needed to better understand the behavior of composite materials in cryogenic conditions and the mutual influence between different materials when exposed to LH_2.

Acknowledgements

S.B., A.M., and I.B. acknowledge the support by the European Union - NextGenerationEU - National Sustainable Mobility Center CN00000023, Italian Ministry of University and Research Decree n. 1033— 17/06/2022, Spoke 12, CUP B73C22000760001. D.C. and I.B. also acknowledge the support of the PON Ricerca e Innovazione 2014-2020 -- Fondo Sociale Europeo, Azione I.1 Dottorati Innovativi con caratterizzazione Industriale -- Ciclo XXXVI (CUP: B73D20005010001 -- Scholarship ID: DOT20KTEXX).

References

[1] Cardone, L.M., Petrone, G., De Rosa, S., Franco, F., Greco, C. S., "Review of the recent development about the hybrid propelled aircraft", *Aerotecnica Missili & Spazio*, 103(1), 17-37, 2024. https://doi.org/10.1007/s42496-023-00173-6

[2] Hoelzen, J., Silberhorn, D., Zill, T., Bensmann, B., and Hanke-Rauschenbach, R., "Hydrogen-powered aviation and its reliance on green hydrogen infrastructure e Review and research gaps," *International Journal of Hydrogen Energy*, 47(5), 3108–3130, 2022. https://doi.org/10.1016/j.ijhydene.2021.10.239

[3] Afonso, F., Sohst, M., Diogo, C. M., Rodrigues, S. S., Ferreira, A., Ribeiro, I., Marques, R., Rego, F. F., Sohouli, A., Portugal-Pereira, J., Policarpo, H., Soares, B., Ferreira, B., Fernandes, E. C., Lau, F., and Suleman, A., "Strategies towards a more sustainable aviation: A systematic review", *Progress in Aerospace Sciences*, 137, 100878, 2023. https://doi.org/10.1016/j.paerosci.2022.100878

[4] Suewatanakul, S., Porcarelli, A., Olsson, A., Grimler, H., Chiche, A., Mariani, R., and Lindbergh, G., "Conceptual Design of a Hybrid Hydrogen Fuel Cell/Battery Blended-Wing-Body Unmanned Aerial Vehicle—An Overview" *Aerospace*, 9(5), 275, 2022. https://doi.org/10.3390/aerospace9050275

[5] Adler, E. J., and Martins, J. R. R. A., "Hydrogen-Powered Aircraft: Fundamental Concepts, Key Technologies, and Environmental Impacts", *Progress in Aerospace Sciences*, 2023. https://doi.org/10.1016/j.paerosci.2023.100922

[6] Brewer, G. D., Hydrogen aircraft technology, Routledge, 1991

[7] Lee, J. A, and Woods, S., "Hydrogen Embrittlement", National Aeronautics and Space Administration Scientific and Technical Information (STI) Program Office, 2016

Aerospace Science and Engineering - IV Aerospace PhD-Days
Materials Research Proceedings 42 (2024) 80-83

Materials Research Forum LLC
https://doi.org/10.21741/9781644903193-18

Coupling VLM and DG methods for the aeroelastic analysis of composite wings

Dario Campagna[1,a], Alberto Milazzo[1,b], Ivano Benedetti[1,c *], Vincenzo Gulizzi[1,d *]

[1]Department of Engineering, Università degli Studi di Palermo, Viale delle Scienze, Edificio 8, 90128 Palermo, Italy

[a]dario.campagna@unipa.it, [b]alberto.milazzo@unipa.it, [c]ivano.benedetti@unipa.it, [d]vincenzo.gulizzi@unipa.it

Keywords: Vortex Lattice Method, Discontinuous Galerkin Method, Divergence Speed, Aeroelasticity, Composite Wing

Abstract. In this work, a novel computational tool is introduced for rapid aeroelastic analysis of composite wings, integrating an aerodynamic Vortex Lattice Method with a structural Interior Penalty Discontinuous Galerkin method. After the mathematical description of the aeroelastic model, some results for a composite wing are presented to investigate the influence of varying lamination angles on the wing displacement, twist, and divergence speed. Validation against commercial software confirms the effectiveness of the approach.

Introduction

The ability to predict how structures respond to aerodynamic forces is essential for various applications in civil, energy, and aerospace engineering. For instance, modern high-altitude long-endurance aircraft feature very flexible high-aspect ratio wings, prone to large deformations; therefore, their design must carefully account for aeroelastic effects, i.e. the mutual interaction between the deformation of the wing structure and the aerodynamic loads acting on it.

Solving the aeroelastic response of a structure is a fluid-structure interaction problem, which generally cannot be solved analytically and requires computational approaches based on the combined use of numerical codes for structural and fluid mechanics [1, 2]. One solution strategy could involve the coupling of three-dimensional elasticity and fluid-dynamics models, which would offer high-resolution results but involve a high computational cost of the simulations, significantly limiting their use during preliminary design stages. A different approach is to introduce suitable assumptions on the structural model (e.g., representing the wing as a beam component) and on the fluid model (e.g., assuming potential flow), which would reduce the fidelity of the obtained results but also significantly accelerate the numerical simulations.

The approach proposed here falls within the latter category, and its novelty consists in the coupling of an Interior Penalty Discontinuous Galerkin (DG) method for beam structures [3, 4] and the Vortex Lattice Method (VLM) for potential-flow aerodynamics [5].

Aeroelastic Model

Structural model. Consider a composite flat-plate wing of half-span L, chord c, thickness ζ and sweep angle Λ referred to an orthogonal global coordinate system $Ox_1x_2x_3$, with the axis x_1 along the chord and the axis x_3 along the wing thickness, Fig.(1). Small strains and linear elastic stress-strain relationship are assumed, so that

$$\gamma = \mathcal{D}u \qquad \sigma = C\gamma \qquad (1)$$

where $u = (u_1, u_2, u_3)^T$ is the vector containing the displacement components, $\gamma = (\gamma_{11}, \gamma_{22}, \gamma_{33}, \gamma_{23}, \gamma_{31}, \gamma_{12})^T$ and $\sigma = (\sigma_{11}, \sigma_{22}, \sigma_{33}, \sigma_{23}, \sigma_{31}, \sigma_{12})^T$ are the vectors containing the strain and stress components in Voigt notation, respectively, $\mathcal{D}$ is the strain-displacement linear

Aerospace Science and Engineering - III Aerospace PhD-Days   Materials Research Forum LLC
Materials Research Proceedings 42 (2024) 80-83   https://doi.org/10.21741/9781644903193-18

differential matrix operator, and C is a 6×6 matrix containing the stiffness coefficients. It is worth noting that the stress-strain relationship is written in the global reference system upon assuming that the material be orthotropic in a local reference system $O\tilde{x}_1\tilde{x}_2\tilde{x}_3$, which is defined such that the axis $\tilde{x}_3$ coincides with the axis x_3 and the axis $\tilde{x}_1$ forms an angle θ with the x_1 axis.

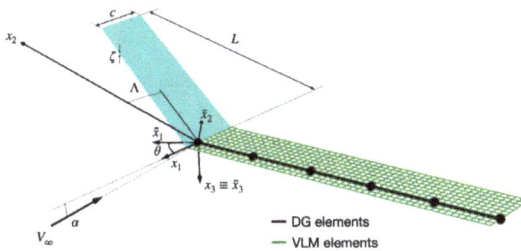

Figure 1. Wing representation.

The wing structure is modelled using high-order generalized beam theories [6, 7, 3] placing the beam elements along the line of quarter-chord points, as sketched in Fig.(1). Following the procedure developed in Ref.[4], it is possible to show the DG weak form of the structural problem reads: find $U^h \in \mathcal{V}^{hp}$ such that

$$B_{\mathrm{DG}}(V, U^h) = L_{\mathrm{DG}}(V, \bar{B}), \qquad \forall V \in \mathcal{V}^{hp}, \tag{2}$$

where $\mathcal{V}^{hp}$ is the space of discontinuous basis functions, U^h is the vector of the generalized displacement components associated with the chosen beam theory and computed using the present DG method, the bilinear form $B_{\mathrm{DG}}(V, U^h)$ accounts for internal elastic energy, and $L_{\mathrm{DG}}(V, \bar{B})$ accounts for the generalized external forces $\bar{B}$ acting on the wing, including the aerodynamic loads. The expression of $B_{\mathrm{DG}}(V, U^h)$ and $L_{\mathrm{DG}}(V, \bar{B})$ can be found in Ref.[4]. It is worth noting that Eq.(2) also includes the kinematic boundary conditions of the wing, which is assumed clamped at its root sections. After numerical integration, computed through standard Gaussian quadrature, the discrete form of Eq.(2) reads as

$$K_S X = F \tag{3}$$

where K_S is the structural stiffness matrix, X collects the coefficients of the DG basis functions, and the vector F collects the associated external forces, which depend on the aerodynamic flow.

Aerodynamic model. The aerodynamic load field is computed using the VLM [5] considering that the lifting surface consists of the geometric surface containing the chord of the wing. Partitioning the lifting surface into a lattice of ring and horseshoe vortices, the VLM framework leads to the following discrete system of equations

$$A\Gamma = b \tag{4}$$

where Γ collects the unknown vortex strengths, and A and b are the aerodynamics coefficient matrix and right-hand side, respectively, stemming from flow impenetrability conditions.

Aeroelastic coupling. The aeroelastic coupling is derived upon noticing that, in general, the right-hand side of Eq.(3) depends on the aerodynamic forces and thus on Γ, while the coefficients matrix A and b depend on the unit normal of the lifting surface and thus on the deformation of the structure. The coupled aeroelastic problem can then written as

$$\begin{cases} A(X)\Gamma = b(X) \\ K_S X = F_A(\Gamma) \end{cases} \tag{5}$$

Following the procedure discussed in Ref.[4], the aeroelastic system given in Eq.(6) can be solved by a linearly coupled analysis, which leads to

$$\left[K_S - \rho_\infty V_\infty^2 \left(\frac{\partial \widehat{F}_A}{\partial \Gamma} A^{-1} \frac{\partial \widehat{b}}{\partial X} \right) \Big|_{\Gamma=0, X=0} \right] X = \left(\frac{\partial F_A}{\partial \Gamma} A^{-1} b \right) \Big|_{\Gamma=0, X=0}, \tag{6}$$

where ρ_∞ is the free-stream density, V_∞ is the free-stream velocity magnitude, while $\widehat{F}_A$ and $\widehat{b}$ coincide with F_A and b, respectively, computed using $\rho_\infty = 1$ and $V_\infty = 1$. The solution of Eq.(6) provides the aeroelastic response of the structure, whereas solving the eigenvalue problem on the left-hand side of Eq.(6) allows computing the divergence speed.

Results

The proposed framework has been implemented using PySCo[1]. The considered case study is an unswept rectangular flat-plate wing, made of an orthotropic material with half-span $L = 10$ m, chord $c = 1$ m and thickness $\zeta = 100$ mm. The material properties in the local material reference system are $E_1 = 20.5$ GPa, $E_2 = E_3 = 10$ GPa, $v_{12} = v_{13} = v_{23} = 0.25$ and $G_{12} = G_{13} = G_{23} = 5$ GPa. Fig.(2) shows the effect of the lamination angle θ on the leading edge vertical displacement u_z and on the twist Δu_z, respectively, of the tip cross-section using free stream velocity $V_\infty = 50$ m/s, angle of attack $\alpha = 1°$ and $\rho_\infty = 1.225$ kg/m^3.

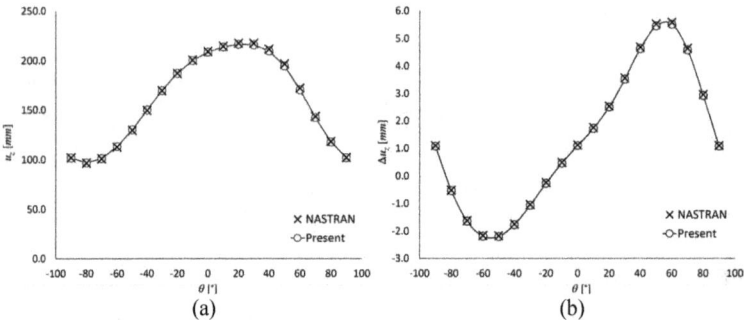

Figure 2. Effect of the lamination angle θ on (a) the leading-edge vertical displacement u_z and (b) on the twist Δu_z of the wing tip cross-section.

The numerical results were obtained using a structural discretization of five eighth-order order DG elements implementing a third-order beam theory, and a VLM lattice composed of 10×50 vortices. The same discretization setup was employed to compute the divergence speed of the wing as a function of the lamination angle θ and three different values of the sweep angle, i.e., $\Lambda = 0°, -10°, -20°$, see Fig.(3). The results were compared with those obtained using the commercial software NASTRAN, demonstrating excellent agreement, and validating the proposed model.

---

[1] https://gitlab.com/aeropa/pysco

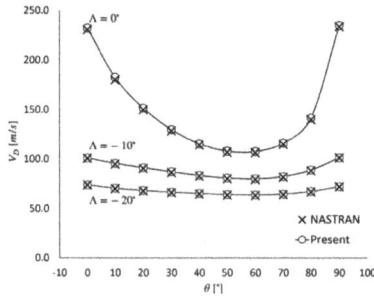

Figure 3. Effect of the lamination angle θ and the sweep angle Λ on the divergence speed V_D.

Conclusions

A computational tool for aeroelastic analysis of composite wings has been proposed. The framework couples an aerodynamic VLM with a structural Interior Penalty DG method for composite wings. The effect of the composite lamination angle on the maximum displacement and twist of the wing tip and on the divergence speed has been studied. The obtained results have been validated by commercial FEM. Further studies will investigate more complex geometries coupled with non-planar VLM approaches and high aspect ratio wings in large-strains.

Acknowledgements

DC and IB acknowledge the support of the PON Ricerca e Innovazione 2014-2020 – Fondo Sociale Europeo, Azione I.1 Dottorati Innovativi con caratterizzazione Industriale – Ciclo XXXVI (CUP: B73D20005010001 – Scholarship ID: DOT20KTEXX). VG acknowledges the support by the European Union – Next Generation EU - PNRR M4 - C2 -investimento 1.1: Fondo per il Programma Nazionale di Ricerca e Progetti di Rilevante Interesse Nazionale (PRIN) - PRIN 2022 cod.2022AALLEC – Project HYDRA CUP B53D23005770006.

References

[1] M. Grifò, V. Gulizzi, A. Milazzo, A. Da Ronch and I. Benedetti, High-fidelity aeroelastic transonic analysis using higher-order structural models. *Composite Structures*, 321:117315, 2023. https://doi.org/10.1016/j.compstruct.2023.117315

[2] M. Grifò, A. Da Ronch and I. Benedetti. A computational aeroelastic framework based on high-order structural models and high-fidelity aerodynamics. *Aerospace Science and Technology*, 132:108069, 2023. https://doi.org/10.1016/j.ast.2022.108069

[3] V. Gulizzi, I. Benedetti and A. Milazzo, High-order accurate beam models based on discontinuous Galerkin methods, *Aerotecnica Missili & Spazio*, 102(4):293-308, 2023. https://doi.org/10.1007/s42496-023-00168-3

[4] V. Gulizzi and I. Benedetti, Computational aeroelastic analysis of wings based on the structural discontinuous Galerkin and aerodynamic vortex lattice methods. *Aerospace Science and Technology*, 144:108808, 2024. https://doi.org/10.1016/j.ast.2023.108808

[5] J. Katz and A. Plotkin, *Low-speed aerodynamics*, (13). Cambridge university press, 2001. https://doi.org/10.1017/CBO9780511810329

[6] E. Carrera, M. Cinefra, M. Petrolo, E. Zappino, *Finite element analysis of structures through unified formulation*. John Wiley & Sons, 2014. https://doi.org/10.1002/9781118536643

[7] L. Demasi, Y. Ashenafi, R. Cavallaro, E. Santarpia, Generalized unified formulation shell element for functionally graded variable-stiffness composite laminates and aeroelastic applications. *Composite Structures*, 131:501–515, 2015. https://doi.org/10.1016/j.compstruct.2015.05.022

Aerospace Science and Engineering - IV Aerospace PhD-Days   Materials Research Forum LLC
Materials Research Proceedings 42 (2024) 84-88   https://doi.org/10.21741/9781644903193-19

Coupled thermoelastic analysis using 1D higher-order structural theories and finite elements

Martina Santori[1,a] *

[1]Mul2 Lab, Department of Mechanical and Aerospace Engineering, Politecnico di Torino, Corso Duca degli Abruzzi 24, 10129 Torino, Italy

[a]martina.santori@polito.it

Keywords: Thermoelasticity, Carrera Unified Formulation, Finite Element Method

Abstract. This paper presents solutions to coupled thermoelastic dynamic problems of beams subjected to thermal loads over time. A higher-order one-dimensional (1D) model in the framework of the Carrera Unified Formulation (CUF) is used. The study aims to provide accurate predictions for displacement fields and temperature changes within homogeneous isotropic structures under thermal loads. A numerical test case describing the influence of sudden heating on the response of beam structure is presented. The results of the quasi-static analysis are compared with the dynamic response. Different two-dimensional Lagrange expansions are used to discretize the beam cross-section. The approach used in this work simplifies the complex three-dimensional (3D) problem into a computationally efficient 1D model.

Introduction

The study of thermoelastic phenomena has gained significant interest in recent years, given its considerable importance in many engineering applications. In the aerospace sector, thermal stress may be significant given the extreme operating conditions components often face, characterized by high temperatures and rapid temperature fluctuations. The complex interplay between thermal effects and mechanical responses has led to the development of numerous models for studying thermoelastic problems. Static thermal loading involves stationary temperature distributions that affect component deformations, while quasi-static loading accounts for time-dependent temperature changes that lead to transient thermal stresses. On the other hand, dynamic thermoelasticity introduces additional complexities, such as inertial effects, as external thermomechanical loads fluctuate rapidly over time.

Traditionally, thermoelastic problems have been addressed using uncoupled theories, in which temperature and mechanical displacements are treated independently. Although these models offer computational simplicity, they may not accurately predict the behavior of aerospace structures under extreme operating conditions. As such, there is a growing need for more sophisticated theories of coupled thermoelasticity.

In 1956, Biot [1] presented the classical theory of thermoelasticity, according to which thermal disturbances propagate with infinite velocity within the structure. Afterwards, other theories, such as the Lord-Shulman (LS) [2] and Green-Lindsay (GL) [3] models, were developed that overcome the limitations of the classical theory.

The advent of numerical methods, particularly the finite element analysis, has enabled the accurate simulation of complex thermoelastic phenomena. This paper uses a refined 1D model based on the Carrera Unified Formulation [4]. This approach has demonstrated the ability to perform multi-field analysis, particularly thermomechanical analysis, on complex structures, such as beams or disks, in an accurate and computationally efficient way [5,6]. This paper shows how the FE-CUF 1D model allows for accurate coupled dynamic analysis using Newmark's implicit method for resolution. Some numerical results on quasi-static and dynamic problems on an isotropic, homogeneous beam subjected to impulsive thermal loading are presented.

Governing equations

The first general governing equation of the coupled thermoelasticity is the equation of motion in a three-dimensional domain [7]:

$$\sigma_{ij,j} + X_i = \rho \ddot{u}_i + \zeta \dot{u}_i \tag{1}$$

where σ_{ij} is the stress component, X_i is the volume forces and u_i is the displacement component. ρ and ζ are the density and damping coefficient, respectively. The derivative in space is defined by subscript (,) while the derivative in time is denoted by superscript ($\cdot$).

The stress component is expressed by Hooke's law for a non-homogeneous anisotropic material as:

$$\sigma_{ij} = C_{ijpq}\epsilon_{pq} - \beta_{ij}(T + t_1\dot{T}) \tag{2}$$

where ϵ_{pq} is the strain component, C_{ijpq} is the 4th-order elasticity tensor, T is the temperature change with respect to the reference temperature T_0 and t_1 is one of the two relaxation times predicted by Green-Lindsay (GL) theory. The parameter β_{ij} è il second-order tensor of termoelastic moduli.

The energy equation can be expressed as a function of displacements u_i and temperature T [7]:

$$\rho c(t_0 + t_2)\ddot{T} + \rho c\dot{T} - 2\tilde{c}_i\dot{T}_{,i} - \left(\kappa_{ij}T_{,j}\right)_{,i} + t_0 T_0 \beta_{ij}\ddot{u}_{i,j} + T_0\beta_{ij}\dot{u}_{i,j} = R + t_0\dot{R} \tag{3}$$

where c is the specific heat, $\tilde{c}$ is a vector of material constants and t_0 and t_2 are the relaxation times relative to Lord-Shulman (LS) theory and Green-Lindsay (GL) theory, respectively. The parameter κ_{ij} is the thermal conductivity tensor.

Eq. 1 and Eq. 3 represent the coupled governing equations of coupled thermoelasticity written in the most general form. To solve the two equations simultaneously, a finite element formulation is adopted using the virtual displacement principle (PVD) [4]:

$$\delta L_{int} = \delta L_{ext} - \delta L_{ine} \tag{4}$$

where δL_{int}, δL_{ext} and δL_{ine} are the internal, external, and inertial virtual works, respectively.

The displacement field $\boldsymbol{u}$ and temperature variation T can be expressed by CUF using the finite element method through the following relations [4]:

$$\boldsymbol{u} = N_m F_\tau \boldsymbol{U}^{m\tau}; \quad T = N_m F_\tau \Theta^{m\tau} \tag{5}$$

where N_m are the shape functions, F_τ are the generic expansion functions and $\boldsymbol{U}^{m\tau}$ and $\Theta^{m\tau}$ are the generalized vector of displacements and the generalized temperature change, respectively. Different Lagrange expansions are used in this work. Using the equations written according to CUF and Hooke and geometric equations within the PVD, the governing equations can be written in the following matrix form:

$$\begin{bmatrix} M_{UU}^{lm\tau s} & 0 \\ M_{\theta U}^{lm\tau s} & M_{\theta\theta}^{lm\tau s} \end{bmatrix} \begin{Bmatrix} \ddot{U}^{ls} \\ \ddot{\theta}^{ls} \end{Bmatrix} + \begin{bmatrix} G_{UU}^{lm\tau s} & G_{U\theta}^{lm\tau s} \\ G_{\theta U}^{lm\tau s} & G_{\theta\theta}^{lm\tau s} \end{bmatrix} \begin{Bmatrix} \dot{U}^{ls} \\ \dot{\theta}^{ls} \end{Bmatrix} + \begin{bmatrix} K_{UU}^{lm\tau s} & K_{U\theta}^{lm\tau s} \\ 0 & K_{\theta\theta}^{lm\tau s} \end{bmatrix} \begin{Bmatrix} U^{ls} \\ \theta^{ls} \end{Bmatrix} = \begin{Bmatrix} F^{ls} \\ Q^{ls} \end{Bmatrix} \tag{6}$$

Aerospace Science and Engineering - III Aerospace PhD-Days   Materials Research Forum LLC
Materials Research Proceedings 42 (2024) 84-88   https://doi.org/10.21741/9781644903193-19

where the terms of the matrices are expressed through a condensed formulation that does not depend on the order of the model, the so-called fundamental nuclei. For brevity, the expressions for the matrix equation terms are not given in this paper but are explicitly presented in [7].

Newmark's method [8] is used for solving Eq. 6, which in compact form is written:

$$\widehat{K} q_{t+\Delta t} = \widehat{R}_{t+\Delta t} \tag{7}$$

where $q_{t+\Delta t}$ is the vector of the unknowns of displacements and temperature, $t + \Delta t$ is the time step, $\widehat{K}$ is the effective stiffness matrix and $\widehat{R}_{t+\Delta t}$ is the effective loads. The method involves solving the system iteratively for each instant of time. The explicit relations of the terms in the equation and more details of the method can be found in [8].

Numerical results

The case examined is an isotropic beam with a square cross-section clamped at one edge. The beam has a dimensionless length of $\bar{L} = 0.5$ and a section edge equal to $\bar{I} = 0.05$. The clamped face is subject to a temperature change $\bar{T}(\bar{x}, \bar{y} = 0, \bar{z}, \bar{t}) = 1 - e^{-100\bar{t}}$, where $\bar{t}$ is the dimensionless time. The previous parameters are given in the dimensionless form given the very small dimensions and times considered in this case. The relationships between dimensionless and dimensional parameters can be found in [5]. The material is aluminum and has the following characteristics: the Lamè constants are $\lambda = 40.4$ GPa, $\mu = 27$ GPa and $\rho = 2707$ kg m^{-3}, $\alpha = 23.1 \cdot 10^{-6}$K^{-1}, $\kappa = 204$ Wm^{-1}K^{-1} [5]. The reference temperature is $T_0 = 293$ K. The model adopted consists of ten 4-nodes finite elements along the y-axis of the beam and different Lagrange elements to model the cross-section, such as 4-node bilinear (1L4), 9-node biquadratic (1L9), and 16-node bicubic elements.

The quasi-static response of the structure is initially analyzed. The time history of the dimensionless temperature and axial displacement in the midpoint of the structure are shown in Fig. 1.

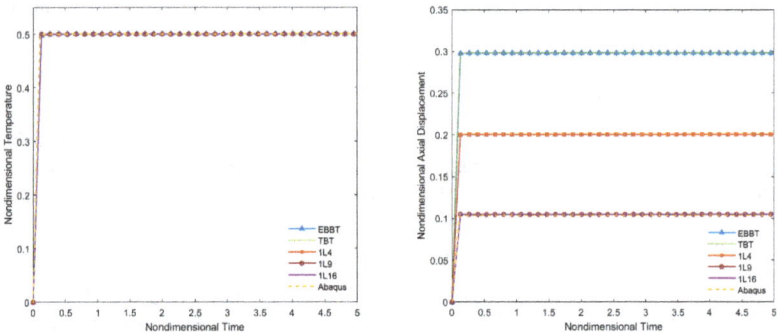

Figure 1: Comparison of nondimensional temperature and axial displacement results of quasi-static analysis obtained through different Lagrange and Abaqus models.

Solutions for different Lagrange models are compared with those obtained from the classical Euler-Bernoulli (EBBT) and Timoshenko (TBT) theories. The results are also compared with the quasi-static response obtained with Abaqus software. The comparison shows that the 1L9 model represents a good trade-off between accuracy and computational efficiency.

In the same time range, the dynamic response using the generalized Lord-Shulman theory is shown in Fig. 2 compared with the quasi-static case. The dimensionless relaxation time of the LS theory is assumed to be $\bar{t} = 0.64$. The inclusion of inertial effects results in fluctuations of temperature and displacement trends around the value obtained from the quasi-static case. In Fig. 2, the dynamic response using the LS model is also compared with the results obtained from the classical dynamic theory. Unlike the classical theory, the LS model manages to predict temperature fluctuations before reaching the steady-state value. Furthermore, the classical theory underestimates the value of the displacement. The trends of temperature and axial displacements are verified through comparison with the solution proposed by Filippi et al. [5] shown in Fig. 2.

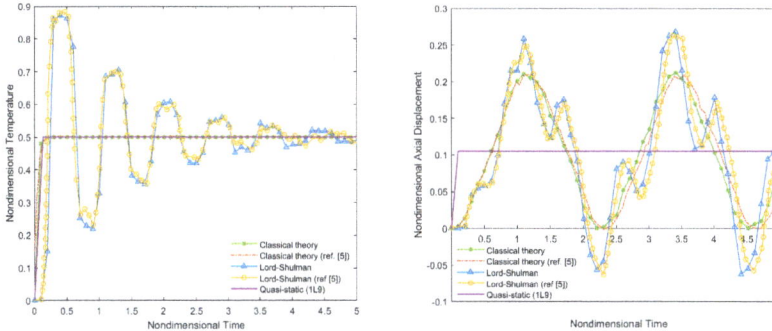

Figure 2: Time history of temperature and axial displacement obtained from dynamic analysis compared with reference results [5] and the quasi-static case.

Summary

This paper presents some numerical results of coupled dynamic thermoelasticity on a homogeneous and isotropic beam. The trends of temperature change and displacements of the structure for the quasi-static and dynamic cases were obtained using a refined 1D model based on the Carrera Unified Formulation (CUF). The approach used was validated through comparison with a commercial code and reference solutions. The results show that by neglecting inertial effects, the quasi-static analysis cannot predict the temperature and displacement fluctuations inferred from the dynamic theory. In addition, using generalized theories such as the Lord-Shulman model allows for a more accurate response of the structure by predicting oscillations in temperature that classical theory could not.

Acknowledgements

This work was partly supported by the Italian Ministry of Foreign Affairs and International Cooperation, grant number US23GR12.

References

[1] M. A. Biot. Thermoelasticity and irreversible thermodynamics. *Journal of applied physics.* 27, 240-253 (1956). https://doi.org/10.1063/1.1722351

[2] H. W. Lord, Y. Shulman. A generalized dynamical theory of thermoelasticity. *Journal of the Mechanics and Physics of Solids.* 15, 299–309, 1967. https://doi.org/10.1016/0022-5096(67)90024-5

[3] A. E. Green, K. A. Lindsay. Thermoelasticity. *Journal of elasticity.* 2, 1–7, (1972). https://doi.org/10.1007/BF00045689

[4] E. Carrera, M. Cinefra, M. Petrolo, E. Zappino. Finite element analysis of structures through unified formulation. *John Wiley & Sons* (2014). https://doi.org/10.1002/9781118536643

[5] M. Filippi, A. Entezari, E. Carrera. Unified finite element approach for generalized coupled thermoelastic analysis of 3D beam-type structures, part 2: Numerical evaluations. *J. Therm. Stresses*. 40.11, 1402-1416 (2017). https://doi.org/10.1080/01495739.2017.1336741

[6] M.A. Kouchakzadeh, A. Entezari, E. Carrera. Exact Solutions for Dynamic and Quasi-Static Thermoelasticity Problems in Rotating Disks. *Aerotec. Missili Spaz.* 95, 3–12 (2016). https://doi.org/10.1007/BF03404709

[7] A. Entezari, M. Filippi, E. Carrera. Unified finite element approach for generalized coupled thermoelastic analysis of 3D beam-type structures, part 1: Equations and formulation. *J. Therm. Stresses*. 40.11,1386-1401 (2017). https://doi.org/10.1080/01495739.2017.1336740

[8] K. J. Bathe. Finite element procedures. Klaus-Jurgen Bathe (2006). https://doi.org/10.1002/9780470050118.ecse159

Aerospace Science and Engineering - IV Aerospace PhD-Days
Materials Research Proceedings 42 (2024) 89-92

Materials Research Forum LLC
https://doi.org/10.21741/9781644903193-20

Hybrid-triggered orbit-attitude coupled station keeping control of a spacecraft with pure thrusters

Hongyi Xie[1,a] * and Franco Bernelli-Zazzera[1,b]

[1]Department of Aerospace Science and Technology, Politecnico di Milano, Via La Masa, 34, 20156, Milano, Italy

[a]hongyi.xie@polimi.it, [b]franco.bernelli@polimi.it

Keywords: Hybrid-Triggered Control, Orbit-Attitude Coupling, Non-Spherical Gravity

Abstract. This paper introduces a specialized attitude-orbit control system designed for three-axis stabilized micro-spacecraft facing the challenge of maintaining orbit radius near small asteroids. Utilizing a fixed main thruster and inclined small thrusters, the system addresses intense orbit-attitude coupling. An intermittent hybrid-triggered control method optimizes main thruster alignment with the asteroid's center of gravity and maintains the spacecraft's orbit within a specified range. The system responds to attitude deviations by activating only the inclined thrusters and triggers overall control for specified orbital conditions, ensuring spacecraft safety. Simulation results confirm the efficacy of the proposed control system.

Introduction

Small spacecraft, like CubeSats, offer cost-effective, rapid, and versatile solutions for deep space exploration. Their small size allows for faster development, technology testing, and collaboration [1]. With lower costs, multiple spacecraft can be deployed, mitigating risks and increasing mission success. Additionally, small spacecraft contribute to educational opportunities and are suitable for interplanetary exploration, enabling missions to asteroids, moons, and planets [2]. However, the actuators in small spacecraft are simpler compared to their larger counterparts. The maximum output of reaction wheels in a 50kg micro spacecraft is always limited to no more than 0.02 Nm [3]. An attitude control system with a reaction wheel might be impractical for a lightweight Cube spacecraft, weighing only 1kg. Consequently, inclined thrusters are consistently incorporated into these kinds of spacecraft to generate sufficient attitude control torques as needed. A classical layout of a small spacecraft with inclined thrusters is given as follows:

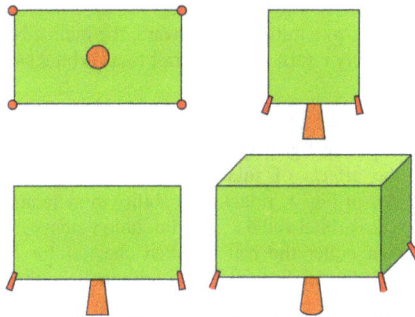

Fig.1. Three view drawing of a small spacecraft with a classical layout of 4+1 thrusters

Aerospace Science and Engineering - III Aerospace PhD-Days   Materials Research Forum LLC
Materials Research Proceedings 42 (2024) 89-92   https://doi.org/10.21741/9781644903193-20

Fig. 1 illustrates a classical layout of thrusters on a small spacecraft. There are 4 inclined thrusters and 1 main thruster fixed on the same side of this spacecraft, where the main thruster is fixed on the center of this surface with its direction orthogonal to this surface of the spacecraft. Compared with the main one, the four inclined thrusters can provide smaller thrusts, with their direction a little deviating from the axis direction of the main thruster at the same small angle. Note that the directions of the 4 main thrusters are very close to that of the main thruster. Therefore, this spacecraft can provide a high control torque in two directions of attitude control, for example roll and pitch, while the maximum theoretical control torque that could be given by this spacecraft is much smaller along the yaw axis. This layout is associated very well with the mission of an orbital station keeping in the vicinity of an asteroid, with a high requirement for control torques in roll and pitch owing to the non-spherical gravity and the gravity gradient and a very low need for control torques in yaw if the inertia matrix of the spacecraft is selected properly [4]. Consequently, integrated orbit and attitude control of a small spacecraft in the vicinity of an asteroid is possible by using this layout with a reasonable design of the control system.

The spacecraft's orbital thruster can operate continuously [5] or intermittently [6]. Continuous thrust may requires a complex inner structure, making intermittent thrusters a cost-effective and more robust choice for small spacecraft. Stabilizing a spacecraft within a narrow range around the center of a celestial body poses challenges, particularly with intermittent thrusters. An intermittent event-triggered control scheme effectively confines the spacecraft within a safety zone, limiting orbital radius and velocity direction. This approach has proven successful in pure orbital control [7] and orbital control considering orbit-attitude coupled dynamics [8] near asteroids. However, dealing with the manipulation of spacecraft attitude using only inclined thrusters requires a different approach. A hybrid-triggered control scheme [9] becomes valuable for executing necessary control orders when the spacecraft's attitude deviates too far or is at risk of leaving the designated mission area.

Therefore, this paper introduces a hybrid-triggered control scheme to effectively confine the orbital radius of a small spacecraft within a defined area. This hybrid-triggered approach integrates an improved version of the intermittent event-triggered orbit control scheme from [8] and another intermittent event-triggered scheme focusing on the spacecraft's attitude states. Each subsection's triggering initiates a new activation moment, employing thrusters to alter the spacecraft's motion states. However, the control response varies significantly depending on the triggering conditions. If the hybrid scheme is triggered by a substantial attitude error, specific inclined thrusters engage to reduce the error. By contrast, if triggered by jeopardized orbital states, the inclined thrusters address pitch and roll attitude errors initially. Afterward, the inclined thrusters deactivate, and the main thruster takes over to safely return the spacecraft to its orbital station.

Problem Formulation

The spacecraft's dynamics model aligns closely with our prior research [8], incorporating considerations for the non-spherical gravity of the asteroid, the significant impact of the gravity gradient around it, and the effects of solar radiation pressure (SRP). However, due to the spacecraft's layout illustrated in Fig. 1, it lacks the ability to generate thrust directed downward, hindering the reduction of its orbital radius. This limitation stems from the necessity of a 180-degree attitude maneuver in either the roll or pitch channel for initiating downward thrust. Executing such a maneuver incurs high energy costs with inclined thrusters. The hybrid-triggered mechanism is designed to activate before the completion of such a maneuver, utilizing the asteroid's gravity for this purpose. Accordingly, the "mid-line" orbital radius, denoted as r_0, is set below the average of the upper and lower limits, ensuring the avoidance of triggering by the orbital subsection in the range $\|r\| \in (r_0, r_{max})$. Here, $\|r\|$ represents the distance from the spacecraft's center to the asteroid's center, and r_{max} corresponds to the upper limit of the safety area's orbital

Aerospace Science and Engineering - III Aerospace PhD-Days   Materials Research Forum LLC
Materials Research Proceedings 42 (2024) 89-92   https://doi.org/10.21741/9781644903193-20

radius. Besides, the layout of the spacecraft and also the special settings about the inclined thrusters are presented in the following figure:

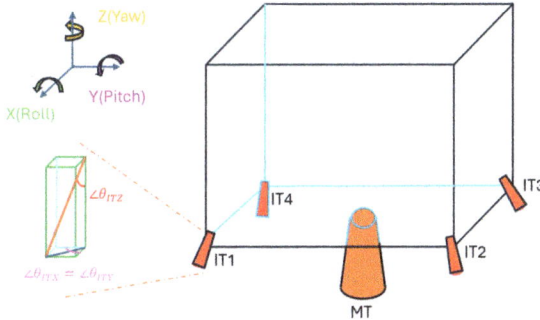

Fig.2. The spacecraft's layout with 4 inclined thrusters and corresponding definitions

The thrust generated by inclined thrusters 1, 2, 3, or 4 is consistently set to a fixed value denoted as F_I, and the inclined angles are uniformly defined with varying orientations, as illustrated in Fig. 2. Achieving 3-axis attitude control involves activating different combinations of inclined thrusters. For instance, rolling about the X-axis is attainable by simultaneously activating IT2 and IT3 or by concurrently engaging IT1 and IT4 for a roll motion in the opposite direction. Pitching about the Y-axis can be realized by activating IT1 and IT2 together or by concurrently opening IT3 and IT4 for pitch motion in the opposite direction. Similarly, yawing about the Z-axis is achievable by simultaneously activating IT1 and IT3 or by concurrently opening IT2 and IT4 for yaw motion in the opposite direction, provided that the thrust direction of IT1 and IT3 or IT2 and IT4 is not on the same plane.

Methodology
The hybrid-triggered mechanism is divided into two subsections. The first subsection is about the attitude control.

$$t_{k1} = \min\{\theta_b \geq b_w\}, k = 1,2,\dots \tag{1}$$

where θ_b denotes the angle between the aimed body-fixed coordinated label system of the spacecraft and the real-time body-fixed coordinated label system of the spacecraft. b_w is the upper limit of the allowed angle of this deviation. Once after this mechanism is triggered, any necessary thruster will be opened to recover the aimed attitude orientation on an almost optimal way. The second subsection is about the orbit control, which is an enhanced version of that in [8]. This subsection is given as follows:

$$t_{k2} = \min\left\{\frac{r_0 - \|\boldsymbol{r}\|}{\|\boldsymbol{r}\|}\boldsymbol{r}\cdot\boldsymbol{v} + \boldsymbol{\alpha} \leq 0\right\}, k = 1,2,\dots \tag{2}$$

where $\boldsymbol{v}$ is the orbital velocity of the spacecraft, and $\boldsymbol{\alpha}$ is a special designed variable to make the triggering condition properly. Similarly, once after this mechanism is triggered, any necessary thruster will be activated to recover the aimed attitude orientation, and then generate enough upwards thrusts to increase the orbital altitude of the spacecraft in an almost optimal way. Combing (1) with (2) yields the hybrid-triggered mechanism as follows:

$$t_k = \min\left\{\theta_b \geq b_w \text{ or } \frac{r_0 - \|\boldsymbol{r}\|}{\|\boldsymbol{r}\|}\boldsymbol{r}\cdot\boldsymbol{v} + \boldsymbol{\alpha} \leq 0\right\}, k = 1,2,\dots \tag{3}$$

Expected Results

Expected results involve effectively constraining the spacecraft's orbital position within the predetermined range. The hybrid-triggered mechanism further limits the spacecraft's attitude deviation within a specified range set by the corresponding subsection. Achieving both orbital and attitude maneuvers requires fewer triggering orders, resulting in a decreased overall thrust requirement from the thrusters. This approach ensures a cost-effective control subsystem for the spacecraft, contributing to an economical solution for the entire system.

References

[1] P. Gaudenzi. "High tech small sat missions with air aunch capabilities." Aerotecnica Missili e Spazio 90 (2011). http://dx.doi.org/10.19249/ams.v90i2-3.62

[2] T. Erdem, S. Speretta, and E. Gill. "Autonomous navigation for deep space small satellites: Scientific and technological advances." Acta Astronautica 193 (2022): 56-74. https://doi.org/10.1016/j.actaastro.2021.12.030

[3] E. Gill, and J. Guo. "MEMS technology for miniaturized space systems: needs, status, and perspectives." In Reliability, Packaging, Testing, and Characterization of MEMS/MOEMS and Nanodevices XI, vol. 8250, pp. 109-118. SPIE, 2012. https://doi.org/10.1117/12.907450

[4] T. Madhur, R. Prazenica, and T. Henderson. "Direct adaptive control of spacecraft near asteroids."Acta Astronautica 202 (2023): 197-213. https://doi.org/10.1016/j.actaastro.2022.10.014

[5] L. De Leo, and M. Pontani. "Low-thrust orbit dynamics and periodic trajectories in the earth–moon system." Aerotecnica Missili & Spazio 101, no. 2 (2022): 171-183. https://doi.org/10.1007/s42496-022-00122-9

[6] Caruso, Andrea, Lorenzo Niccolai, Alessandro A. Quarta, and Giovanni Mengali. "Envelopes of spacecraft trajectories with a single impulse." *Aerotecnica Missili & Spazio* 98 (2019): 293-299. https://doi.org/10.1007/s42496-019-00026-1

[7] P. Ong, G. Bahati, and A. D. Ames. "Stability and safety through event-triggered intermittent control with application to spacecraft orbit stabilization." In *IEEE 61st Conference on Decision and Control*, pp. 453-460. IEEE, 2022. https://doi.org/10.1109/CDC51059.2022.9992757

[8] H. Xie, and F. Bernelli Zazzera. "6-DOF Orbit-attitude Stabilization around an Asteroid by using Performance-based Intermittent Event-triggered Control." In *Aerospace Europe Conference 2023-Joint 10th EUCASS-9th CEAS Conference*, pp. 1-13. 2023. https://doi.org/10.13009/EUCASS2023-027

[9] K. Zhang, and B. Gharesifard. "Hybrid event-triggered and impulsive control for time-delay systems." Nonlinear Analysis: Hybrid Systems 43 (2021): 101109. https://doi.org/10.1016/j.nahs.2021.101109

Aerospace Science and Engineering - IV Aerospace PhD-Days
Materials Research Proceedings 42 (2024) 93-97

Materials Research Forum LLC
https://doi.org/10.21741/9781644903193-21

In situ space debris inspection: From observation to 3D reconstruction

Luca Lion[1,a] *, Chantal Cappelletti [2,b], Samanta Piano[3,c], Francesco Branz[4,d] and Alessandro Francesconi[4,e]

[1]Centre of Studies and Activities for Space (CISAS) "Giuseppe Colombo", University of Padova, via Venezia, 15, Padova, 35131, Italy

[2]Department of Mechanical, Materials and Manufacturing Engineering, University of Nottingham, United Kingdom

[3]Manufacturing Metrology Team, Faculty of Engineering, University of Nottingham, United Kingdom

[4]Department of Industrial Engineering, University of Padova, Via Venezia, 1, Padova, 35131, Italy

[a]luca.lion.1@phd.unipd.it, [b]chantal.cappelletti@nottingham.ac.uk, [c]samanta.piano@nottingham.ac.uk, [d]francesco.branz@unipd.it, [e]alessandro.francesconi@unipd.it

Keywords: On Orbit Servicing, Space Debris, Inspection, Computer Vision

Abstract. Amidst the exponential increase in space debris, Active Debris Removal and On-Orbit Servicing missions have gained paramount importance. The success of such endeavours pivots on the comprehension of the target's geometry and dynamic conditions, highlighting the indispensable nature of target inspection. In recent times, a variety of inspection missions have employed satellites of different classes. In this context, small satellites like CubeSats have emerged as a reasonable solution, due to cost-effectiveness and rapid development capabilities. This study explores a potential inspection mission utilizing a CubeSat to evaluate a non-cooperative target, where information on relative pose are not known a priori. The aim is to retrieve essential data for executing Close Proximity Operations safely. Manoeuvring around the target, the CubeSat captures two-dimensional (2D) RGB images from multiple angles to then reconstruct its three-dimensional (3D) geometry. Optical cameras are preferred during the inspection phase due to their cost-efficiency and low power requirements, different from other technologies like laser imaging. An experimental setup is designed and built to generate a dataset of 2D images simulating in orbit conditions. Standard computer vision algorithms are employed to perform the 3D reconstruction and Artificial Intelligence used to reconstruct the scene.

Introduction

The escalating number of space debris poses a significant threat to the space industry, putting operational satellites and human space missions at risks. Notable incidents like the 2009 collision between Iridium-33 and a Russian military satellite underscore the urgency of addressing the current resident space object (RSO) situation [1, 2]. This has fuelled interest and demand for On-Orbit Servicing (OOS) missions and Active Debris Removal (ADR) applications. During the years, both manned and robotic OOS missions were carried out. For the latter, recent examples are MEV-1 and MEV-2 [3], and the future ESA mission ClearSpace-1 [4]. However, most of these missions targeted cooperative RSOs capable of controlling their attitude, ensuring safer close-proximity operations [5]. When dealing with non-cooperative targets such as tumbling debris, close-proximity operations become riskier due to uncertainties in pose, shape, and conditions. Therefore, in situ inspection of the target is critical to OOS mission success. Previous research has

Aerospace Science and Engineering - III Aerospace PhD-Days   Materials Research Forum LLC
Materials Research Proceedings 42 (2024) 93-97   https://doi.org/10.21741/9781644903193-21

identified potential capture points and fragile components on the targets but falls short in ensuring safe trajectories during close-proximity manoeuvres. To mitigate this risk, understanding the 3D characteristics of the target is essential. Additionally, determining the target's relative pose, crucial for robotic arm capture, requires a 3D model of the target. This study investigates the feasibility of using photogrammetry to inspect an RSO. In addition, the same dataset is used to reconstruct the scene using a variation of NeRF, Neuro Radiance Fields [6]. A custom experimental setup is designed to simulate a CubeSat chaser's fly-around relative to a target RSO, CubeSats are chosen for their fast and relatively easy design process [7]. The collected data are processed to retrieve a 3D model of the target and reconstruct the scene. Preliminary results are then presented and briefly discussed.

Case study

This study focuses on Space Rider (SR), the ESA reusable spacecraft scheduled for a 2025 flight. Structurally, the spacecraft resembles a cylinder, with a diameter of approximately 2 meters and a length of 8 meters as depicted in Figure 2. Positioned near its base are two deployable solar panels, extending to around 11 meters upon deployment. These dimensions classify it as a substantial Resident Space Object (RSO) in Low Earth Orbit (LEO). Moreover, its distinct shape and colour scheme pose significant challenges for 3D model reconstruction techniques. For consistency, the spacecraft's white segment is referred to as the upper part, while the black portion is the lower part. The darker conical structure, housing the solar panels, is known as AVUM +, serving as the upper stage of the Vega-C launcher. To conduct the inspection, a circular relative orbit has been chosen. The inspector, or chaser, will maintain a constant distance of 100 meters from the target, SR.

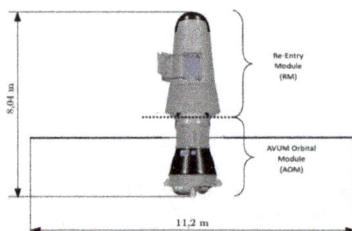

Figure 2: Space Rider dimensions [9]

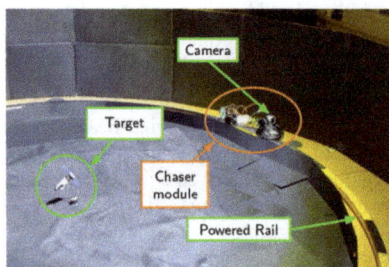

Figure 1: Experimental setup. The chaser is free to move along the rail maintaining a constant distance from the target.

Experimental setup

To replicate the case study in a laboratory setting, a scaled-down setup was designed and built as in Figure 2. Primarily constructed through 3D printing, this setup aims to modularity and ease of

Aerospace Science and Engineering - III Aerospace PhD-Days   Materials Research Forum LLC
Materials Research Proceedings 42 (2024) 93-97   https://doi.org/10.21741/9781644903193-21

reproduction in any laboratory environment. The monorail approach was chosen to provide power to the chaser module. The circular rail has a diameter of 2 meters and beneath it, a gear facilitates the movement of the chaser module. The base of the chaser module mirrors the dimensions of a 3U CubeSat, measuring approximately 300mm × 100mm. Equipped with a machine vision camera featuring a 16mm lens, the chaser is controlled by an onboard Raspberry Pi 4, overseeing both the camera and the rail's motion via a stepper motor. To simulate orbital lighting conditions, the room is obscured and the setup enveloped with matte black panels. The sole source of illumination directed towards the target is a Sun Simulator, emitting a nominal luminous flux of 9000 lumens. The target comprises a 3D-printed, scaled-down 1:100 model of SR.

3D reconstruction techniques

3D scanning techniques are utilized to capture information about an object's shape or other attributes. These techniques find extensive application across various fields in the industry. For instance, they are employed in industrial settings for quality control along production lines and in the entertainment industry to digitally recreate actors and props. Several approaches exist to achieve the desired outcomes, with the most prevalent ones used in space being photogrammetry and LiDAR. The decision for this study leaned towards photogrammetry due to its simplicity and minimal requirements. Unlike LiDAR, photogrammetry does not necessitate specialized instruments, relying solely on a camera and adequate lighting, while also imposing a relatively low computational load. The algorithm employed in this work is the industry-standard Structure from Motion, which, based on features extracted from each image, is capable of determining the relative pose of the cameras.

For reconstructing the scene, Instant-NGP, Neural Graphics Primitives, is chosen as the most viable option. Instant-NGP is a modified version of NeRF created to obtain a faster and less computational heavy network. These characteristics are critical for a future implementation on board of a satellite.

Data processing and results

The pictures are acquired by the chaser during a full lap around the target. In each round, a total of 37 pictures is taken, spaced evenly at 10-degree intervals, with the first and last images overlapping. The camera's position is then estimated using COLMAP, a general-purpose Structure-from-Motion pipeline [8]. The pipeline's general workflow is depicted in Figure 3. Initially, the camera's position is determined, locating them in 3D space. Following this, a sparse point cloud is generated. These initial tasks can be executed using CPU power alone, enabling direct onboard processing using the Raspberry Pi. The last step is the retrieval of the dense point cloud, that can be then used to create the mesh. The final step involves generating a dense point cloud, which can then be used to construct a mesh. To assess the accuracy of the reconstruction, the point cloud is compared to the original CAD model using Cloud Compare, a 3D point cloud processing software. The distance between each element of the point cloud and the nearest point on the mesh is calculated to gauge the quality of the acquired 3D information. The results indicate

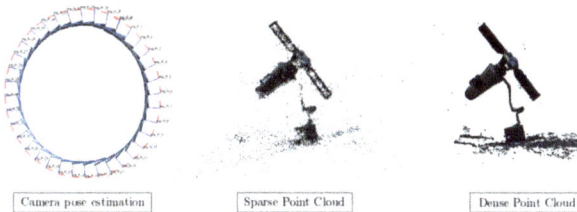

| Camera pose estimation | Sparse Point Cloud | Dense Point Cloud |

Figure 3: 3D reconstruction pipeline.

Aerospace Science and Engineering - III Aerospace PhD-Days   Materials Research Forum LLC
Materials Research Proceedings 42 (2024) 93-97   https://doi.org/10.21741/9781644903193-21

that the mean distance between points was 0.02 mm, with a standard deviation of 0.08 mm, demonstrating the precision of the reconstruction.

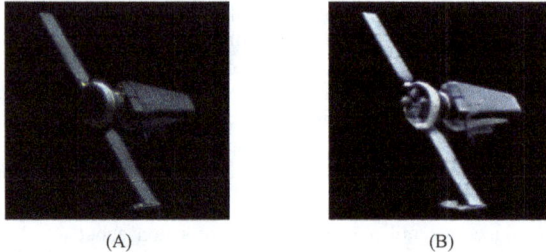

(A)   (B)

Figure 4: On the left (A) the original image, on the right (B) an image from the reconstructed scene.

Regarding Instant-NGP, first the pictures are fed to COLMAP which is used only for the first step retrieving the camera pose. When finished, the dataset and the camera pose file are fed into the neural network which is able to fully reconstruct the scene in less than 2 minutes. For this study an Nvidia RTX 4080 gpu is used. The results show a PSNR of 18.3, a comparison between the original image and the reconstructed one is shown in Figure 4.

Conclusions

In this paper, a preliminary study on the inspection of a Resident Space Object to retrieve its 3D model is presented. The target taken in consideration is Space Rider, an ESA spacecraft designated for operations in Low Earth Orbit (LEO). The goal of the study is to create a 3D model of SR using standard photogrammetry algorithm. Opting for photogrammetry implies that the equipment needed on board on the inspector satellite, specifically a CubeSat in this case, is limited to a camera, a visible RGB for these tests. A custom setup was designed built to simulate the in-orbit conditions. It includes a self-sufficient module, approximately the size of a 3U CubeSat, that orbits the target in a circular path at a fixed distance of 1 meter. The module's camera captures images at regular intervals. The target is a 3D-printed model of ESA Space Rider at a 1:100 scale. After collecting the images COLMAP is used to retrieve the camera pose and generate a point cloud of the environment. This point cloud is then compared to the original CAD model of the target, revealing an average distance of around 0.02 millimetres between the points and the original model. The same camera pose used to then recreate the point cloud is also used as input to a Neural Network, Instant-NGP, to reconstruct the scene. While the findings are still in their early stages, they suggest that this method is suitable for creating a detailed 3D model of the target. Such data could be invaluable for informing the design of future On-Orbit Servicing (OOS) or Active Debris Removal (ADR) missions.

References

[1] N. Johnson, "Orbital debris: the growing threat to space operations," in *33rd Annual Guidance and Control Conference*, 2010.

[2] L. Olivieri, C. Giacomuzzo, S. Lopresti and A. Francesconi, "Simulation of In-Space Fragmentation Events," *Aerotecnica Missili & Spazio*, 2023. https://doi.org/10.1007/s42496-023-00186-1

[3] N. T. Redd, "Bringing satellites back from the dead: Mission extension vehicles give defunct spacecraft a new lease on life-[News]," *IEEE Spectrum* , 2020. https://doi.org/10.1109/MSPEC.2020.9150540

[4] R. Biesbroek, "The clearspace-1 mission: ESA and clearspace team up to remove debris," in *8th Eur. Conf. Sp. Debris*, 2021.

[5] A. Siena, "Orbit/Attitude Control for Rendezvous and Docking at the Herschel Space Observatory," *Aerotecnica Missili & Spazio,* 2023. https://doi.org/10.1007/s42496-023-00188-z

[6] B. Mildenhall, S. P. Pratul, T. Matthew, B. T. Jonathan, R. Ramamoorthi and R. Ng, NeRF: Representing Scenes as Neural Radiance Fields for View Synthesis, ECCV, 2020. https://doi.org/10.1007/978-3-030-58452-8_24

[7] M. Mozzato , M. Bemporad, S. Enzo, F. Filippini, R. Lazzaro, M. Minato, D. Visentin, A. Dalla Via, A. Farina, E. Pilone, F. Basana, L. Olivieri, G. Colombatti and A. Francesconi , "Concept and Feasibility Analysis of the Alba Cubesat Mission," *Aerotecnica Missili & Spazio.*

[8] J.-M. Frahm and J. Schonberger, "Structurefrom-motion revisited," in *IEEE Conference on Computer Vision and Pattern Recognition (CVPR)*, 2016. https://doi.org/10.1109/CVPR.2016.445

[9] S. R. Team, "SPACE RIDER USER GUIDE," 2023.

Aerospace Science and Engineering - IV Aerospace PhD-Days   Materials Research Forum LLC
Materials Research Proceedings 42 (2024) 98-103   https://doi.org/10.21741/9781644903193-22

Integration of a low-cost camera system for smart agriculture aboard tethered balloons and drones

Federico Toson[1,a] *

[1]CISAS G. Colombo – University of Padova, via Venezia, 15 - Italy

[a]federico.toson@phd.unipd.it

Keywords: Smart Agriculture, Remote Sensing, Crop Performance, Balloons, Drones

Abstract. The impacts of climate change on crops are intensifying and notably severe. This considering both natural disasters and the unpredictability of seasonal patterns, which lead to dwindling resources. Therefore, innovative methods are being sought to investigate plant health to reduce waste, intensify productivity, and ensure production continuity. This to date is feasible using satellite and Earth observation technologies in general; however, these pose some challenges for both individual farmers and local organizations. In fact, the cost of high-definition satellite images for surveys of individual crop fields can have very high costs that would not guarantee frequent analysis. On the other hand, other satellite observation strategies, which are cheaper, do not guarantee high ground resolutions and compromise, again, the success of the analysis. For the reasons just listed, a variety of alternative technologies have emerged that, integrated aboard Unmanned Air Vehicles (UAVs), guarantee high-resolution imaging for costs two orders of magnitude lower than satellite imaging. In this essay its described the potential of these devices and provide a brief description of the analysis conducted on crop fields considering their integration on board three different UAVs.

Introduction

Smart agriculture is an ever-increasing growth sector [1]. This is for various reasons, starting from increased production performance, reduced waste and environmental impact, and guarantees provided to end users when purchasing the product.

In recent years, the aerospace sector has focused on developing monitoring solutions that fall under the branch of Earth observation; these technologies and services, offered by governmental [2] and commercial [3] entities, include satellites capable of working in different wavelength of the visible, near-infrared and UV (Ultra Violet) spectrum. In fact, it is not new how, by means of multispectral analyses, a great deal of additional information can be obtained in the observation of nature. Satellite monitoring provides significant contributions from aspects of geology [4], water resources [5], pollution [6] and finally the health of humans [7] and vegetation [8].

However, this revolution is not without its limitations: just as it's impractical to saturate space with satellites, maintaining sporadic observations with satellite passages spaced four or more days apart isn't feasible. In addition, the boundary variables must not be forgotten, such as the atmosphere cloudiness that restricts the view on the ground, the still high cost for a precision satellite services, and the ground resolution that, in many cases, is not sufficient for analyses limited to narrow FOVs (Field of View).

Looking at the monitoring of specific crop fields, the individual farmer, but also a governmental verification body, may find it difficult to obtain repeated quantitative responses over an area of interest.

For these reasons, satellite data require integration with other types of observation technologies that are compact, easy to deploy, compatible with existing technologies, ready to use and easily accessible from an integration and possibly economic perspective.

Aerospace Science and Engineering - III Aerospace PhD-Days   Materials Research Forum LLC
Materials Research Proceedings 42 (2024) 98-103   https://doi.org/10.21741/9781644903193-22

The ATEMO (Aerospace Technologies for Earth Monitoring and Observation) project [9] aims to design and develop these technologies for integration on board mid-atmosphere vehicles such as drones, tethered balloons, and stratospheric balloons, which we will summarise as UAVs.

As part of this project, research activities have been defined that currently focus on the analysis of crop fields. This article will therefore describe these observations from the technology developed, to the on-board integration of UAVs and the innovative contribution to smart agriculture.

Description of analyses

Observations, including satellite monitoring, conducted in the agricultural field are, as mentioned earlier, multispectral. The combination of light information on various spectral bands provides indicators of vegetation health based on the principle that healthy plants emit or absorb different wavelengths than diseased plants. Various examples of this type of analysis can be found in the literature and have become an excellent ally for farmers and agricultural associations [10]. The best-known example is the NDVI (Normalised Difference Vegetation Index), an index that considers the red band at 695 nm, which is related to the absorption of chlorophyll by leaves, and the NIR (Near InfraRed) band 750 nm, which corresponds to the reflectivity of plants. These values are combined into a dimensionless coefficient that gives an indication of the health of the plant under examination and is summarised in the simple formula below (Eq. 1).

$$NDVI = \frac{Red-NIR}{Red+NIR} \tag{1}$$

In addition to this, various indices exist such as the GNDVI (Green Normalised Difference Vegetation Index), ENDVI (Enhanced Normalised Difference Vegetation Index) and many others summarised in official databases shared between researchers and scholars [11]. The diversity among these indices serves a dual purpose. It enables the determination of tailored coefficients for the analysis at hand, while also facilitating the identification of correlations between them when performed concurrently. Consequently, even with limited data, these correlations can be indirectly established. This inherent potential underlies the technologies discussed in this article. By hypothetically conducting high-resolution NDVI assessments on a cultivated field, scientists can validate less resolved satellite data and establish a baseline for monitoring that area, while still relying on technologies affected by certain limitations. However, they can address some of these obstacles and minimize others.

Based on this principle, however, integrating spectrometers on board a drone or tethered balloon is not a viable and cost-effective solution; for this reason, commercial cameras are selected, and, with appropriate optical filtering, wavelengths of interest are isolated. This ensures that almost all the bands seen from the satellite are compensated for and allows for a timely comparison.

In the specific case of ATEMO, an attempt was made to still provide the calculation of at least two or three indices simultaneously, and therefore two cameras were used: one monochrome and one colour.

The colour camera, with triple-band filtering can, on its own, still allow the isolation of three bands; in fact, if distinct spectrum bands are selected and corresponding to the RGB channels of the camera, it is possible, with appropriate calibration of the vision system, to define distinct information on the three selected wavelengths ranges.

The monochrome camera, on the other hand, allows the acquisition of the single reference band corresponding to the filter inserted.

In the case of ATEMO, different filtering solutions can be integrated in this manner, and thus multiple vegetation indices can be defined; below, in Table 1, is a comparison between the bands available from a commercial satellite solution (Planet) and the filters that can be integrated on ATEMO.

Table1: Comparison of the acquisition bands of Planet satellites and ATEMO.

Band name	Planet		ATEMO	
	Nominal Value	Bandwidth	Nominal Value	Bandwidth
Coastal Blue	441,5 nm	21 nm	470 nm	20 nm
Blue	490 nm	50 nm	475 nm	20 nm
Green I	531 nm	36 nm	550 nm	20 nm
Green	565 nm	36 nm		
Yellow	610 nm	20 nm	-	-
Red	665 nm	30 nm	695 nm	20 nm
RE (Red-Edge)	705 nm	16 nm	735 nm	20 nm
Infrared (>800)	-	-	800 nm	-
NIR (Near-Infrared)	865 nm	40 nm	850 nm	20 nm

The most used configuration during the initial system test campaigns was a triple-band 475nm, 550nm and 850nm filter on the colour camera and an RE (Red-Edge) bandpass filter. This is because the focus was on estimating the ENDVI and GNDVI mentioned above, and the NDRE (Normalised Difference Red-Edge index), the definition of which is indicated below (Eq. 2, Eq. 3 and Eq. 4).

$$ENDVI = \frac{(NIR+Green)-2*Blue}{(NIR+Green)+2*Blue} \tag{2}$$

$$GNDVI = \frac{NIR-Green}{NIR+Green} \tag{3}$$

$$NDRE = \frac{NIR-RE}{NIR+RE} \tag{4}$$

The above-mentioned indices are excellent for defining water stress, the subject of analyses conducted during the summer of 2023, however, subsequent filtering options will be tested, especially, those involving blue and near-infrared wavelengths that provide indications of diseases caused by pests and other pathogens that reduce regular plant activity.

Integration and Testing
The total ATEMO system, visible in Figure 1, has been tested aboard tethered balloons, however, reduced versions have been accommodated aboard drones and stratospheric balloons.

Low altitude configurations require high aperture optics, while on balloons reaching high altitudes (between 30 and 40 km) it is essential to change the aperture of the on-board optics.

For integration on-board tethered balloons, continuous analysis spanning the entire day is feasible, ensuring the accumulation of a substantial dataset that captures daily trends effectively. This entails a series of expedients in the definition of the payload such as energy autonomy, acquisition autonomy and the guarantee that, even in strong gusts of wind, the pointing of the cameras remains as stable as possible.

Figure 1: *ATEMO system.*

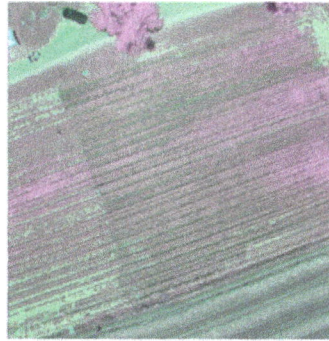

Figure 2: *Field test in Cortona (Arezzo), summer 2023: on the left the image of the monochrome camera with NIR filter and on the right the image of the colour camera with triple-band filters.*

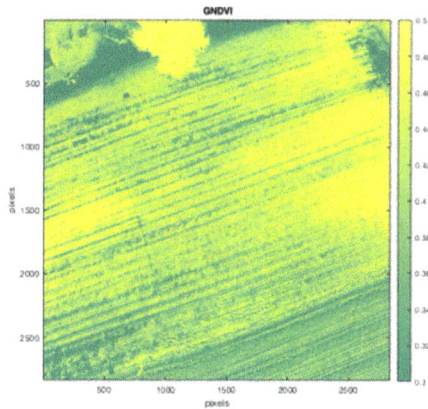

Figure 2: *Example of GNDVI index extrapolation from the photos in Figure 2.*

This is not the case with integration on-board drones which, although they can make multiple passes during a day, have a limited flight time. In this case, the adaptability of the acquisition system is also considered, which for geometric reasons will have to have an angle of inclination to consider when processing the data but above all when defining the flight itinerary. In fact, if the tethered balloon remains stationary at altitudes between 50 and 100 m, the drone flies over the area of interest keeping lower, guaranteeing more defined images but needing to move over the field to cover the entire surface under examination.

In the case of the stratospheric balloon, as mentioned earlier, narrower optical systems are used, and less focused analysis is performed. The lightness of the payload imposed by the flight leads to a reduction in the number of instruments on board, for example, in a test campaign conducted in Pisa, only a single camera (the colour one) was mounted with a triple-band filter.

The testing of the device actively involved the research team and provided several comparable images indicative of its correct functioning and support in the determination of vegetative indices. Above, in Figure 2, is an example of a shot taken by the two cameras with the tethered balloon; while Figure 3 shows an example of reworking to calculate a vegetation index.

Conclusions
In conclusion, ATEMO proves to be a low-cost, versatile, and useful solution for the effects induced by adverse climatic events on crops. During the first test phase, estimates of vegetation indices such as ENDVI, GNDVI and NDRE were conducted with different autonomous flight systems. The tethered balloon proved to be very useful as it allows long autonomous samplings by granting a better manageability of the vehicle by the operator, the drone is more effective for the very accurate analysis of plants and to have resolutions of the order of magnitude of a leaf dimensions. Finally, tests from a tethered balloon, with a system similar to that on the ground, open the way for future comparisons for the validation of satellite data.

The uses of devices such as the one described are varied; soon, the aim is to extend the spectrum of analysis, for example by zooming in on indices to study vines potentially attacked by pests and prevent their spread. In general, the contribution of these technologies is and will be an important resource for fighting climate change and increasing data for analysis and prevention.

Acknowledgements
We would like to thank those who collaborated in the realisation of ATEMO and its testing. Special thanks go to Professor Maurizio Borin and those involved in the PRIN (Project of Significant National Interest) called 'Rewatering'.

References

[1] R. K. Goel, C. S. Yadav, S. Vishnoi and R. Rastogi, "Smart agriculture – Urgent need of the day in developing countries," *Sustainable Computing: Informatics and Systems,* vol. 30, 2021. https://doi.org/10.1016/j.suscom.2021.100512

[2] "Sentinel Online," ESA, [Online]. Available: https://sentinels.copernicus.eu.

[3] Planet, "Daily Earth Data to See Change and Make Better Decisions," Planet, 2023. [Online]. Available: https://www.planet.com.

[4] M. Massironi, L. Bertoldi, P. Calafa, D. Visonà, A. Bistacchi, C. Giardino and A. Schiavo, "Interpretation and processing of ASTER data for geological mapping and granitoids detection in the Saghro massif (eastern Anti-Atlas, Morocco)," *Geosphere,* vol. 4, pp. 736-759, 2008. https://doi.org/10.1130/GES00161.1

[5] G. Raimondi, C. Maucieri, M. Borin, J. L. Pancorbo, M. Cabrera and M. Quemada, "Satellite imagery and modeling contribute understanding cover crop effect on nitrogen

dynamics and water availability," *Agronomy for Sustainable Development,* vol. 43, 2023. https://doi.org/10.1007/s13593-023-00922-8

[6] F. Toson, M. Pulice, M. Furiato, M. Pavan, S. Sandon, D. Sandu and R. Giovanni, "Launch of an Innovative Air Pollutant Sampler up to 27,000 Metres Using a Stratospheric Balloon," *Aerotecnica Missili & Spazio,* vol. 102, no. 2, pp. 127-138, June 2023. https://doi.org/10.1007/s42496-023-00151-y

[7] F. Facchinelli, S. E. Pappalardo, D. Codato, A. Dianini, G. Della Fera, E. Crescini and M. De Marchi, "Unburnable and Unleakable Carbon in Western Amazon: Using VIIRS Nightfire Data to Map Gas Flaring and Policy Compliance in the Yasuní Biosphere Reserve," *Sustainability,* 2020. https://doi.org/10.3390/su12010058

[8] F. Morbidini, W. Barrera Jr., G. Zanin, L. Verdi, C. Canarotto, G. Ghinassi, C. Maucieri, A. Dalla Marta and M. Borin, *The state of the art on deficit irrigation on soybean,* Wiley, 2023. https://doi.org/10.1002/ird.2903

[9] F. Toson, A. Aboudan, C. Bettanini, G. Colombatti, I. Terlizzi, S. Chiodini and L. Olivieri, "The ATEMO device: a compact solution for earth monitoring," in *Aeronautics and Astronautics, AIDAA XXVII International Congress*, Padova, 2023. https://doi.org/10.21741/9781644902813-127

[10] F. Morari, V. Zanella, L. Sartori, G. Visioli, P. Berzaghi and G. Mosca, "Optimising durum wheat cultivation in North Italy: understanding the effects of site-specific fertilization on yield and protein content," *Precision Agriculture,* vol. 19, pp. 257-277, 2017. https://doi.org/10.1007/s11119-017-9515-8

[11] "Index Database," 2024. [Online]. Available: https://www.indexdatabase.de.

Aerospace Science and Engineering - IV Aerospace PhD-Days   Materials Research Forum LLC
Materials Research Proceedings 42 (2024) 104-107   https://doi.org/10.21741/9781644903193-23

High-fidelity simulation and low-order analysis for planetary descent investigation of capsule-parachute interaction

L. Placco[1*], F. Dalla Barba[2] and F. Picano[1,2]

[1]Centro di Ateneo di Studi e Attività Spaziali 'Giuseppe Colombo' (CISAS), Università degli Studi di Padova, via Venezia 15, 35131 Padua, Italy

[2]Department of Industrial Engineering, Università degli Studi di Padova, via Venezia 1, 35131 Padua, Italy

*luca.placco@unipd.it

Keywords: Supersonic Parachute, Supersonic Flows, Large Eddy Simulation, Low-Order Modelling

Abstract. The project focuses on characterizing the unsteady dynamics of the parachute-capsule system during the descent phase of planetary entry in a supersonic flow regime. Currently, Large-Eddy Simulation, coupled with an Immersed-Boundary Method, is utilized to examine the time-evolving flow behavior of a rigid supersonic parachute trailing behind a reentry capsule as it descends through the Martian atmosphere. The flow is simulated at Ma=2 and Re=10^6. A massive GPU parallelization has been utilized to enable a high-fidelity resolution of the turbulent structures in the flow, essential for capturing its dynamic behavior. We demonstrate through low-order modeling of the unsteady turbulent wake of the capsule that low-frequency fluctuations within the wake are the primary trigger for flow instability in front of the canopy volume. Proper-Orthogonal Decomposition is utilized to investigate the system dynamics and analyze how various turbulence contributions influence the phenomenon.

Introduction

The failure of the ExoMars 2016 mission highlighted the persistent difficulties in forecasting and comprehending the dynamics of descent capsules affected by supersonic decelerators. Specifically, the unsuccessful landing of the Schiaparelli Entry, Descent, and Landing Module (EDM) was ultimately linked to an inadequate assessment of the interconnected oscillations between the descent module and the deployed parachute. Models and experimental assessments utilized to anticipate the capsule's response to the supersonic decelerator's impact were deemed inadequate, resulting in the mission's premature termination [1].

The inherent non-linearity of both fluid and solid behavior poses a significant challenge when studying unsteady compressible flows, particularly in the context of interactions between turbulent flows and solid structures [2,3]. This complexity is further compounded in scenarios involving unsteady compressible flows. Thus, the primary objective of this research is to develop effective approaches to investigate the interaction between compressible and turbulent flows, specifically focusing on the unsteady effects observed during flight [4]. The study relies on high-fidelity simulations of fluid dynamics, aiming to minimize the introduction of models and disturbances to provide the most accurate representation of flow dynamics under expected flight conditions. Currently, the analysis excludes the representation of flexible structures to concentrate solely on fluid-dependent effects and highlight the root causes of the observed phenomenon. Results from the baseline simulation are further analyzed to identify the primary trigger of the parachute bow shock instability, primarily attributed to low-frequency disturbances introduced by the wake of the descent module. This study builds upon previous analyses of descent module wake frequency conducted by the same authors [5]. Additionally, Proper-Orthogonal Decomposition is employed

Aerospace Science and Engineering - III Aerospace PhD-Days   Materials Research Forum LLC
Materials Research Proceedings 42 (2024) 104-107   https://doi.org/10.21741/9781644903193-23

to gain deeper insights into turbulence dynamics, directly correlating frequency contents with oscillation patterns in the flow field that surrounds the canopy.

Computational setup and approach

Compressible Navier-Stokes equations are solved using the high-order finite difference solver STREAmS [6]. Turbulent structures are simulated through the implicit Large-Eddy Simulation (ILES) approach, eliminating conventional LES turbulence modeling. Instead, the numerical discretization provides artificial viscosity acting at small scales, effectively capturing turbulent behavior. Thus, the three-dimensional compressible Navier-Stokes equations solved are the following:

$$\frac{\partial \rho}{\partial t} + \frac{\partial (\rho u_j)}{\partial x_j} = 0.$$

$$\frac{\partial (\rho u_i)}{\partial t} + \frac{\partial (\rho u_i u_j)}{\partial x_j} + \frac{\partial p}{\partial x_i} - \frac{\partial}{\partial x_j}\left(\mu \left(\frac{\partial u_i}{\partial x_j} + \frac{\partial u_j}{\partial x_i} - \frac{2}{3}\frac{\partial u_k}{\partial x_k}\delta_{ij} \right) \right) = 0.$$

$$\frac{\partial (\rho E)}{\partial t} + \frac{\partial (\rho E u_j + p u_j)}{\partial x_j} + \frac{\partial}{\partial x_j}\left(\lambda \frac{\partial T}{\partial x_j} \right) + \frac{\partial}{\partial x_j}\left(\mu \left(\frac{\partial u_i}{\partial x_j} + \frac{\partial u_j}{\partial x_i} - \frac{2}{3}\frac{\partial u_k}{\partial x_k}\delta_{ij} \right) u_i \right) = 0.$$

where ρ is the density, u_i denotes the velocity component in the i Cartesian direction (i =1,2,3) and p is the thermodynamic pressure. With the intent of reproducing the effect of Mars' atmosphere, the fluid is considered as an ideal gas of CO_2; the ratio between the specific heat at constant pressure C_p and the specific heat at constant volume C_v is set to 1.3 while Prandtl number is 0.72. $E = C_v T + u_i^2/2$ represents the total energy per unit mass and the dynamic viscosity μ is assumed to follow the generalized fluid power-law. The thermal conductivity λ is related to μ via the Prandtl number with the following expression: $\lambda = C_p \mu/Pr$. The simulation is conducted at $Ma = 2$ and $Re = 10^6$ to replicate the conditions during parachute deployment. The reference fluid properties corresponding to the free-stream condition are obtained at an altitude of approximately 9 km from the planet's surface, simulated through an entry and descent trajectory in the Mars atmosphere of a generic reentry probe [7]. The flow domain selected to perform this first simulation has a size of $L_x = 20D$, $L_y = 5D$, $L_z = 5D$, where D is the maximum diameter of the descent module; parachute diameter is set to 2.57D. the mesh is a rectilinear structured grid that consists of $N_x \cdot N_y \cdot N_z = 2560 \cdot 840 \cdot 840$ nodes.

Computations have been carried out on CINECA Marconi100 cluster, computing on 64 GPUs.

Modal analysis and more specifically Proper-Orthogonal Decomposition (POD) has been applied to extract orthogonal basis which contribute to form a linear composition of the original flow field $q(x,y,z,t)$ as follows [8]:

$$q(x, y, z, t) - \overline{q}(x, y, z) = \sum_j a_j(t)\phi_j(x, y, z)$$

where $\overline{q}$ represent the time-averaged flow obtained from q and φ_j and a_j are respectively the spatial mode and the temporal coefficients which drives them. Flow modes are sorted by their respective energy content: the first extracted are the most energetic and the last ones are the least energetic: further details can be found in [5].

Results and outcomes

In figure 1 we observe the three-dimensional representation of the flow around the descent module and its interaction with the parachute via the use of Q-Criterion method; capsule and parachute bow shock are visible. The Q-Criterion technique reveals turbulent structures in the wake of the

Aerospace Science and Engineering - III Aerospace PhD-Days   Materials Research Forum LLC
Materials Research Proceedings 42 (2024) 104-107   https://doi.org/10.21741/9781644903193-23

module as they propagate downstream towards the decelerator. Near the position of the parachute bow shock, eddies originating from upstream experience a notable increase in size, forming larger vortices as they traverse the compression discontinuity. This transitional effect acts as the primary driver of the unsteady dynamics surrounding the decelerator in supersonic regimes. Turbulent fluctuations flowing downstream from the wake induce localized disturbances in the parachute shock speed, disrupting its equilibrium with the upstream flow and initiating continuous cyclic motion of the shock wave. This motion correlates with periodic variations in flow pressure within the canopy volume, ultimately influencing parachute performance. By employing low-order decomposition, we are able to further study the turbulence dynamics based on its energy contents and relative pattern of oscillations. In Figure 2, energy contours of selected flow modes in the region surrounding the capsule wake and the parachute are depicted. It is observed that the first modes, representing the most energetic fluctuations, correspond to the largest oscillation patterns and carry the lowest frequency content in the flow turbulence. Within these oscillation patterns, the canopy bow shock is distinctly visible at the first modes, exhibiting a disrupted shape due to the effect of unsteadiness. Notably, the canopy bow shock completely disappears at higher modes, indicating a direct relationship between shock motion and instability caused by the large fluctuations originating from the capsule wake. Specifically, the low-frequency, high-energy contents of the incoming turbulence contribute to the oscillations that propagate through the shock and toward the canopy. Based on these considerations, a zero-dimensional simplified model is built to describe the unsteady dynamics of shock motion and decelerator performance. Providing as an input only the lower frequency content of the wake dynamics, we observe comparable oscillations in both the canopy flow state and also the drag performance of the parachute.

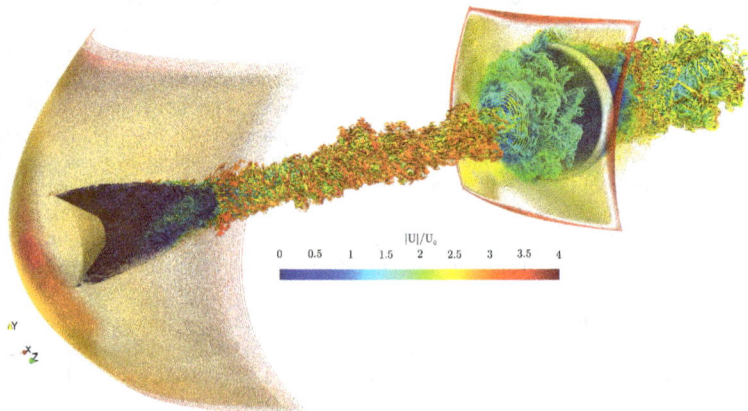

Figure 1*: Q-criterion 3D representation of the flow around the descent module and parachute using non-dimensional velocity magnitude.*

Materials Research Forum LLC
https://doi.org/10.21741/9781644903193-23

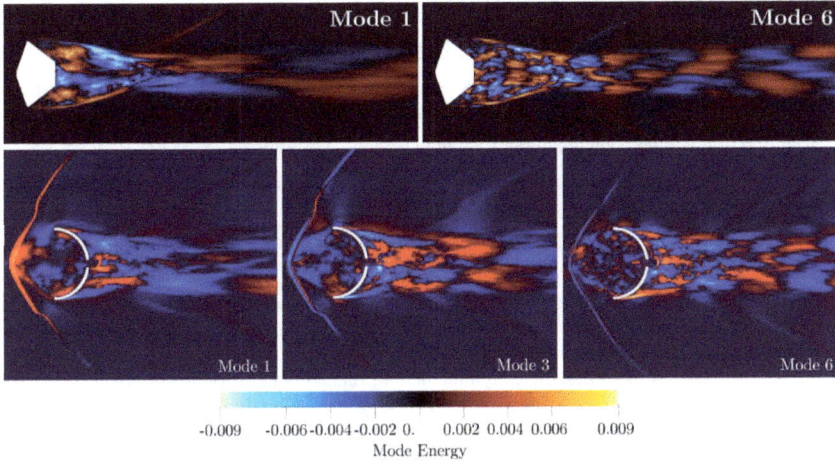

Figure 2: Energy contours for selected POD flow modes ($y/D = 0$ cross section, crop views). Descent capsule, top; parachute canopy, bottom.

References

[1] T. Tolker-Nielsen. EXOMARS 2016 - Schiaparelli Anomaly Inquiry, 2017.

[2] X. Xue and Chih-Yung Wen. Review of unsteady aerodynamics of supersonic parachutes. Progress in Aerospace Sciences, 125:100728, 2021. ISSN 0376-0421. https://doi.org/10.1016/j.paerosci.2021.100728

[3] Nimesh, Dahal. Study of pressure oscillations in supersonic parachute. International Journal of Aeronautical Space Sciences, (19):24–31, 2018. https://doi.org/10.1007/s42405-018-0025-3

[4] B. S. Sonneveldt, I. G. Clark, and C. O'Farrell. Summary of the Advanced Supersonic Parachute Inflation Research Experiments (ASPIRE) Sounding Rocket Tests with a Disk-Gap-Band Parachute, AIAA 2019-3482. https://doi.org/10.2514/6.2019-3482

[5] L. Placco, M. Cogo, M. Bernardini, A. Aboudan, F. Ferri and F. Picano. Large-Eddy Simulation of the unsteady supersonic flow around a Mars entry capsule at different angles of attack. Aerospace Science and Technology, 143:108709, 2023. https://doi.org/10.1016/j.ast.2023.108709

[6] M. Bernardini, D. Modesti, F. Salvadore, and S. Pirozzoli. Streams: a high-fidelity accelerated solver for direct numerical simulation of compressible turbulent flows. Computer Physics Communications, 263:107906, 2021. https://doi.org/10.1016/j.cpc.2021.107906

[7] A. Aboudan, G. Colombatti, C. Bettanini, F. Ferri, S. Lewis, B. Van Hove, O. Karatekin, and Stefano Debei. Exomars 2016 schiaparelli module trajectory and atmospheric profiles reconstruction. Space Science Reviews, 214: 97, 08 2018. https://doi.org/10.1007/s11214-018-0532-3

[8] K. Taira, et al., Modal analysis of fluid flows: an overview, AIAA J. 55 (2017) 4013–4041. https://doi.org/10.2514/1.J056060

Aerospace Science and Engineering - IV Aerospace PhD-Days

Materials Research Forum LLC

Materials Research Proceedings 42 (2024) 108-112

https://doi.org/10.21741/9781644903193-24

Active debris removal employing a robotic arm equipped CubeSat

Federico Basana[1,a] *, Francesco Branz[2,b] and Alessandro Francesconi[2,c]

[1]Centre of Studies and Activities for Space "Giuseppe Colombo" (CISAS), University of Padova, via Venezia, 15, 35131 Padova, Italy

[2]Dip. Ing. Industriale, Università degli Studi di Padova, via Venezia, 1, 35131 Padova, Italy

[a]federico.basana@phd.unipd.it, [b]francesco.branz@unipd.it, [c]alessandro.francesconi@unipd.it

Keywords: In Orbit Servicing, Active Debris Removal, CubeSat, Robotic Arm

Abstract. Space debris pose significant risks to functioning satellites. To mitigate the issue, the space sector is studying new technologies for Active Debris Removal (ADR) missions. Conventional methods rely on large and complex satellites equipped with robotic arms but the high economic cost of these may outweigh the benefit related with the debris removal. This paper proposes a novel, cost-effective approach. A 12U CubeSat equipped with a robotic arm is employed to attach an Elementary Servicing Unit (ESU) to target satellites. The CubeSat integrates essential subsystems such as power management, an Attitude and Orbit Control System (AOCS), a mono-ocular camera for navigation, and a four-degrees-of-freedom robotic arm. To study the feasibility of the mission, a simulator in the MATLAB/Simulink environment has been developed together with a guidance, navigation and control (GNC) system. In addition, a mock-up of the proposed CubeSat has been developed for testing a simple manoeuvre in relevant laboratory environment to attach the ESU to a target. The preliminary results obtained from the simulation and the design of the CubeSat mock-up are presented in the paper.

Introduction

Space debris are a risk for satellites in near-Earth orbits [1]. Their number is expected to grow in the next future due to the creation of large constellation and the proposed and launched small satellites in crowded orbits [2, 3]. Specifically, space debris are a danger since they may collide with operative satellites generating the Kessler's syndrome. To address this issue, satellites and space debris are constantly monitored in the context of the Space Surveillance and Tracking (SST) [4] and Space Situational Awareness [5]. If a risk of collision arises, the operative satellite has to perform a Collision Avoidance Manoeuvre [6] that results in a shortening of the mission lifetime. In addition, the definition of guidelines [7] has regulated the end-of-life procedures of a satellite with the aim of reducing the risk related with space debris. Specifically, in [7] the long-term effectiveness of some debris mitigation measures is studied: the removing of massive objects from the most densely populated orbits may result in a long-term stability of the space debris population. Hence, the interest in space missions with the goal of autonomously capture and remove a space debris has grown in the last decade. Several GNC strategies have been proposed [8] and some demonstration missions have been proposed, e.g., e.Deorbit [9], or performed as the Mission Extension Vehicle 1 (MEV-1) and Mission Extension Vehicle 2 (MEV-2) [10]. In general, these employ large and complex satellites equipped with robotic arms to perform the task required by the Active Debris Removal (ADR) mission. The cost related to the former approach are high and may offset the benefits seen by the operator related to the removal of the debris. Moreover, ADR missions require Close Proximity Operations (CPOs) between two satellites, hence there exists the risk of collision between the two.

This study introduces an innovative approach to satellite removal. A CubeSat sized satellite is used to perform the ADR tasks instead of a large and complex satellite. The main advantage of

Aerospace Science and Engineering - III Aerospace PhD-Days   Materials Research Forum LLC
Materials Research Proceedings 42 (2024) 108-112   https://doi.org/10.21741/9781644903193-24

exploiting CubeSat system is that they are less expensive than large satellites and their development is faster [11]. In addition, the effects of an unforeseen collision between the servicer and the target are less severe under the space debris generation point of view according to EVOLVE 4.0 breakup model [12]. This work studies the feasibility of a CubeSat-sized servicer for ADR missions. Specifically, it presents the development of a mock-up for testing servicing manoeuvres in representative laboratory conditions. Together with the design of the mock-up, the paper shows the analyses performed in the MATLAB/Simulink environment to validate the guidance, navigation and control system of the mock-up.

Mission description

The mission has the objective to deorbit a small satellite in LEO. In the scenario, the satellite, called target, is a 100 kg platform representative of the ones used by large constellations such as OneWeb or Starlink. The target is considered prepared for servicing: it has fiducial markers, e.g., ArUco markers, that aid the navigation algorithm of the servicer and it has an interface for being captured by another satellite. In addition, it is able to maintain its attitude during the capture manoeuvre. The satellite that perform the ADR operations, called servicer, is a 12U CubeSat equipped with a 4 Degree of Freedom (DoF) robotic arm. Besides the subsystems required for power and data generation and distribution, the servicer has an Attitude and Orbit Determination and Control system (AOCS) made of reaction wheels and thrusters that permit orbit and attitude manoeuvres. In addition, it features a navigation system based on a monocular camera for navigation. The robotic arm is used to manipulate and attach a Elementary Servicing Unit (ESU) to the target. The ESU is a 1U sized CubeSat that contains a drag augmentation device designed for deorbiting LEO satellites such as the Commercial Off The Shelf (COTS) drag sail "ARTICA CubeSat Deorbiting System". Once attached, the drag sail contained in the ESU will reduce the deorbiting time of the target exploiting the atmospheric drag. The mission is divided in several phases: (i) orbit transfer and phasing, (ii) far– and mid– range rendezvous, (iii) inspection, (iv) target approach, (v) target capture (vi) deorbiting. In the study only the fourth and fifth phases are considered. These two phases are the most critical under the GNC point of view since the servicer operates in the close proximity of the target.

The servicer mock-up

To study the feasibility of the proposed mission, a 12U CubeSat mock-up of the servicer satellite has been developed (Figure 1) [13]. The aim of the mock-up is to simulate simplified close proximity operations manoeuvres in a representative laboratory environment, i.e., a low-friction table. The mock-up employs a pressurized air system that features three air bearings that allow the

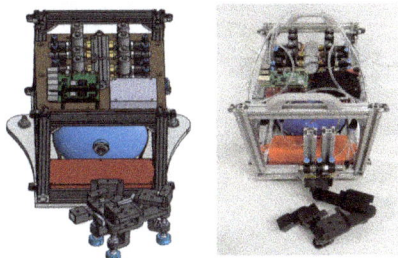

Figure 1: CAD representation of the mock-up (left) and developed mock-up (right)

floating of the module. In addition, 16 nozzles, connected to 8 electrovalves, permit the planar motion of the mock-up. A 2.5 L tank contains the pressurized air (10 bar) required for feed both the air bearing and the thrusters. The main computer of the mock-up is a Raspberry Pi which has the function to control the module and the robotic arm. In addition, another Raspberry Pi is used for the navigation purposes i.e., image acquisition and processing. The latter board uses the CAN bus protocol to pass to the main computer the information concerning the relative pose of the mock-up with respect to the target. The robotic arm is mounted on a face of the mock-up and it is composed of four revolute joints, thus having four degrees of freedom. An air bearing is mounted under each joint to avoid the bending of the robotic arm caused by gravity. The joints are composed by DC motors connected to Hall-effect encoders that ensures the knowledge of the joints position. The end-effector is simplified as an electromagnet that permit to handle a mock-up of the ESU.

The GNC algorithm and results
Different GNC algorithms can be used to control the robotic arm and the module. The free-floating strategy controls the motion of the end-effector keeping the base uncontrolled. In this fashion the reaction torques created by the manipulator, which are disturbances for the mock-up attitude, are not controlled and are free to affect the attitude of the mock-up. This method is used if the mass of the satellite is much greater than the mass of the manipulator so that the disturbances are negligible. Another strategy is called free-flying, in which two control plants work in parallel to maintain the pose of the satellite and move the end effector respectively. In this fashion the disturbances generated by the manipulator are balanced by the AOCS of the satellite. In both the free-floating and free-flying approaches, the redundancy of the manipulator can be leveraged to reduce the disturbance torques as shown in [14]. An alternative is the combined control strategy, where the satellite actuators and the robotic arm joints are seen as multiple degrees of freedom of the same control plant. Hence, the controller can employ the thrusters at the same time as the robotic arm is extending to grab the target [15].

A simulator has been developed in the MATLAB/Simulink environment to simulate the manoeuvre. In this study the free-floating method is applied. During the target approach (which starts 0.6 m distant from the target) the robotic arm is kept in the folded configuration and a PID controller commands the thruster to reach a hold point 0.2 m distant from the grasping point. The pose information is provided to the GNC system as the ideal value (measured by Simulink) plus a white noise to simulate the real behaviour of a sensor. Then, the control start moving the end-effector to reach the grasping point while leaving the satellite body free to evolve under the disturbance torques generated by the manipulator motion. During the manoeuvre, the guidance defines the pose of the end effector in the cartesian space. Using the inverse kinematics of

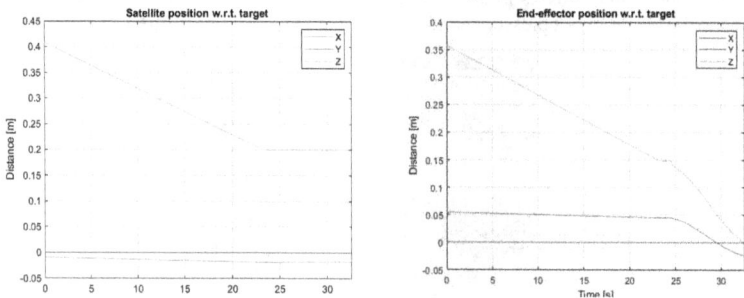

Figure 2: Satellite position w.r.t. the target (left), end-effector position w.r.t. the target (right).

redundant manipulators, the commanded joint positions are retrieved. Then, the commanded signal is compared with the value measured by the encoders and it is provided to a PI controller to obtain the voltage required for the motor actuation.

Aerospace Science and Engineering - III Aerospace PhD-Days   Materials Research Forum LLC
Materials Research Proceedings 42 (2024) 108-112   https://doi.org/10.21741/9781644903193-24

Figure 2 shows the simulation results. In the first 22 s the mock up approaches the target. Then, it waits 1 s and starts moving the robotic arm to bring the end effector to the grasping point. The end effector reaches the target with an error in module lower than 0.02 m. In addition, from these preliminary results it can be seen how an error in the satellite base is reflected to the end effector position during the motion.

Conclusions

In this study a novel approach to the Active Debris Removal is proposed. Specifically, it has been studied the feasibility of using a CubeSat sized servicer to attach a drag sail to a satellite in LEO. A mock-up of the CubeSat has been developed for testing purposes in relevant laboratory environment. A free-floating GNC algorithm has been developed and tested in the MATLAB/Simulink environment. The preliminary results shown the good performances of the algorithm and the importance of a robust and precise control during CPOs with a robotic arm. Future works will investigate other control strategies both in simulation and in laboratory environment.

References

[1] T. Maclay, D. Mcknight, Space environment management: Framing the objective and setting priorities for controlling orbital debris risk, Journal of Space Safety Engineering 8 (1) (2021) 93–97. https://doi.org/10.1016/j.jsse.2020.11.002

[2] H. G. Lewis, G. G. Swinerd, R. J. Newland, The space debris environment: future evolution, The Aeronautical Journal 115 (1166) (2011) 241–247. https://doi.org/10.1017/S0001924000005698

[3] L. Olivieri, A. Francesconi, Large constellations assessment and optimization in leo space debris environment, Advances in Space Research 65 (1) (2020) 351–363. https://doi.org/10.1016/j.asr.2019.09.048

[4] P. Faucher, R. Peldszus, A. Gravier, Operational space surveillance and tracking in europe, Journal of Space Safety Engineering 7 (3) (2020) 420–425. https://doi.org/10.1016/j.jsse.2020.07.005

[5] M. Polkowska, Space situational awareness (ssa) for providing safety and security in outer space: implementation challenges for europe, Space policy 51 (2020). https://doi.org/10.1016/j.spacepol.2019.101347

[6] Z. Pavanello, et al., A Convex Optimization Method for Multiple Encounters Collision Avoidance Maneuvers, AIAA SciTech Forum, Orlando, Florida, (2024). https://doi.org/10.2514/6.2024-0845

[7] H. Klinkrad, P. Beltrami, S. Hauptmann, C. Martin, H. Sdunnus, H. Stokes, R. Walker, J. Wilkinson, The esa space debris mitigation handbook 2002, Advances in Space Research 34 (5) (2004) 1251 – 1259. https://doi.org/10.1016/j.asr.2003.01.018

[8] Stolfi, A., Gasbarri, P. & Sabatini, M. Performance Analysis and Gains Tuning Procedure for a Controlled Space Manipulator Used for Non-cooperative Target Capture Operations. Aerotecnica Missili & Spazio. 97, 3–12 (2018). https://doi.org/10.1007/BF03404759

[9] S. Estable, et al.: Capturing and deorbiting envisat with an airbus spacetug. results from the esa e.deorbit consolidation phase study. Journal of Space Safety Engineering 7(1), 52–66 (2020). https://doi.org/10.1016/j.jsse.2020.01.003

[10] C. Gohd, A northrop grumman robot successfully docked to a satellite to extend its life. URL https://www.space.com/northrop-grumman-mev-2-\docks-intelsat-satellite

[11] M. Mozzato, M. Bemporad, S. Enzo, F. Filippini, R. Lazzaro, M. Minato, D. Visentin, A. Dalla Via, A. Farina, E. Pilone, F. Basana, L. Olivieri, G. Colombatti and A. Francesconi, "Concept and Feasibility Analysis of the Alba Cubesat Mission," Aerotecnica Missili & Spazio.

[12] N.L. Johnson, P.H. Krisko, J.-C. Liou, P.D. Anz-Meador, NASA's new breakup model of evolve 4.0, Advances in Space Research, Volume 28, Issue 9, 2001,1377-1384, https://doi.org/10.1016/S0273-1177(01)00423-9

[13] S. Galleani, T. Berthod, A. Caon, L. Lion, F. Basana, L. Olivieri, F. Branz, A. Francesconi, Mechanical and pneumatic design and testing of a floating module for zero-gravity motion simulation, XXVII AIDAA Congress, 4-7 september 2023, Padova, Italy. https://doi.org/10.21741/9781644902813-118

[14] F. Basana, F. Branz: Simulation of robotic space operations with minimum base reaction manipulator. Journal of Space Safety Engineering 9 (2022). https://doi.org/10.1016/j.jsse.2022.06.005

[15] Z. Pavanello, et al.: Combined control and navigation approach to the robotic capture of space vehicles. In: 72nd International Astronautical Congress (IAC) (2021)

Aerospace Science and Engineering - IV Aerospace PhD-Days
Materials Research Proceedings 42 (2024) 113-116

Materials Research Forum LLC
https://doi.org/10.21741/9781644903193-25

Feedback control of a density-based space debris model to support the definition of efficient mitigation and remediation strategies

Martina Rusconi[1,a] *, Lorenzo Giudici[1,b], Camilla Colombo[1,c]

[1]Politecnico di Milano, Via La Masa 34, 20256, Milano, Italy

[a]martina.rusconi@polimi.it, [b]lorenzo1.giudici@polimi.it, [c]camilla.colombo@polimi.it

Keywords: Space Debris, Feedback Control, Sustainability, Density Propagation

Abstract. Space can be seen as an extension of our planet's biosphere, and as happens with Earth's ecosystem, humanity's utilisation of it is not sustainable. The number of in-orbit debris is dramatically growing and the current guidelines for limiting their proliferation are less adequate to face nowadays situation. The research proposes to develop a systematic way to investigate the effectiveness of the different mitigation and remediation measures. This is done integrating a density-based model for in-orbit objects propagation with a feedback controller on the environment, mimicking human actions in space, to reach a target scenario. The tool will be a valuable support to the definition of a new strategy for the sustainable future utilisation of space.

Introduction

Life on Earth dramatically relies on space assets, but future access to orbit is threatened by the uncontrolled exploitation of space. When scientist Donald Kessler in the '70s proposed his theory on the proliferation of debris in space, less than 5000 objects were tracked orbiting Earth. In the following decades that number has grown exponentially and today hundreds of thousands of debris as little as 1 mm are estimated to pollute our planet's orbital space [1]. In 2002 the Inter-Agency Debris Committee (IADC) formulated guidelines to limit the proliferation of debris [2], that focussed on disposal of spent bodies, reduction of explosions risk and collisions prevention. However, recently, miniaturisation of technologies and a more affordable access to orbit fuelled a new concept of space economy and the consequent increase in launch traffic [3]. According to the future predictions of the European Space Agency's (ESA) Annual Environment Report the current utilisation of space is not sustainable [1]. Therefore, it is now mandatory to define new and updated guidelines suited to face the fast-evolving situation, such as the recent ESA's Zero Debris Policy to limit creation of debris in ESA's future missions [4]. There is broad agreement in this context that it is required to synergistically combine wider adoption of mitigation measures and innovative remediation techniques [1][5]. As of today, simulators for space objects evolution have been used to investigate the effect of specific counteractions. ESA's DELTA software was exploited in [6] to characterise the possible future debris scenarios by manually changing the requirements of IADC guidelines, and in [3] the effect of different static levels of Post-Mission Disposal (PMD) adherence were analysed. However, there is still no consensus on how to efficiently combine these elements in a suitable strategy towards a sustainable space utilisation. A fast and systematic approach is necessary to investigate the predicted effectiveness of the many possible debris mitigation actions.

To address this need, the research proposes the integration of an evolutionary model of the space population with an active feedback controller. Mitigation and remediation measures enter the system as control inputs, mimicking human actions in space and on mission design, as shown in Fig. 1. The control logic tunes the selected inputs in time acting on the objects' distribution in space to reach a target scenario. The approach allows for fast and versatile analyses of the many possible futures predicted by the model under the effect of diversified rules.

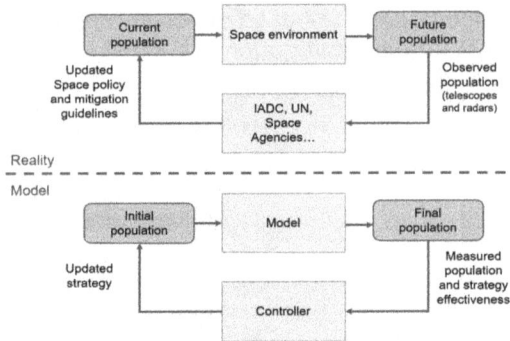

Fig. 1. Schematic of the model (bottom) compared to the real-world interactions it aims to represent (top). The figure is inspired from the one in [10].

Methodology

Fig. 1 shows the main blocks of the tool that is being developed within the research. The space objects' propagator exploits the approach of describing the clouds of debris as a flow with continuous properties. Leveraging on previous works, such as [7][8][9], the dynamic is defined by enforcing mass conservation through Eq. (1), in which the time evolution of the density n is affected by the dynamics f and the deposition and removal rates $\dot{n}^+, \dot{n}^-$.

$$\frac{\partial n}{\partial t} + \nabla \cdot (nf) = \dot{n}^+ - \dot{n}^- \tag{1}$$

The applicability and effectiveness of such density-based methods have been extensively proven in the literature and advanced multi-dimensional tools for debris environment propagation through continuum mechanics have been developed in recent years, such as the Starling and COMETA suites [9][10]. However, a simplified one-dimensional model is first considered, to privilege rapidity of the analyses on accuracy in this preliminary development phase. Referring to the work in [7][8], Eq. (1) is formulated in terms of orbital radius only. The spatial domain is divided in spherical shells and the evolution of the objects' density is captured through the finite volume method, adapting the analysis in [9] to the one-dimensional case. Eq. (2) shows the system of equations resulting by applying the divergence theorem to Eq. (1) of each ith shell within its limiting radii $r_{i_{lower}}, r_{i_{upper}}$, in which the dynamic term v_r considers only the natural effect of atmospheric drag, exploiting the simplified exponential model for air density in [7].

$$\begin{cases} \dot{n}_i = \frac{1}{V_i}\left(-\left(4\pi n_{i+1}v_{r_{i+1}}r_{i_{upper}}^2 - 4\pi n_i v_{r_i}r_{i_{lower}}^2\right) + \int_V (\dot{n}_i^+ - \dot{n}_i^-)dV\right) \\ \vdots \end{cases} \tag{2}$$

Expanding the work in [8], source and sink mechanisms included in $\dot{n}$ account for new launches, objects moved through PMD and active debris removals. Each of them is modelled as a continuous contribution in time and space to the density rate. These functions can take any shape to mimic real behaviours, such as historical profiles, or to investigate alternatives in space utilisation. Additionally, in-orbit fragmentations will be included based on the collision and explosion probabilities of the objects. The first feedback control logic analysed is the quadratic proportional one proposed in [11] for its simple definition and physical interpretation. It is reported in Eq. (3), in which the gain proportional to the squared error e^2 is the ratio of maximum control allowed u_{max} with the squared maximum error e_{max}^2 between the shell's density n_i with respect to a

Aerospace Science and Engineering - III Aerospace PhD-Days   Materials Research Forum LLC
Materials Research Proceedings 42 (2024) 113-116   https://doi.org/10.21741/9781644903193-25

reference value n_{ref}, above which the control is saturated. For the preliminary results shown in the following the error is defined as in Eq. (3).

$$u = \frac{u_{max}}{e_{max}^2} e^2 \qquad \text{where} \qquad e = n_{ref} - n_i \tag{3}$$

In future works, more complex definitions will be considered, such as PID or linear quadratic controllers, and robustness will be analysed accounting for uncertainty. Any source or sink term in Eq. (1) can be considered, in principle, as a control input. Up to now, both launches and PMD models have been included in the propagator and the feedback controller can act on the objects' distribution in space and time by changing the launches deposition rate $\dot{n}_{launches}$ or the compliance factor λ that scales the PMD contribution as $\lambda \dot{n}_{PMD}$ to reach a predefined target.

The tool will benefit from its extreme versatility to characterise many future scenarios aiming at the most promising strategies to reach sustainability of the environment. Combinations of the control inputs will be considered and diversified rules analysed in time for different species of objects and different orbital regions.

Preliminary results

Since the research is still in its preliminary phase, the results of a simple example are provided in the following considering a scenario that is not a realistic representation of the debris environment but has the sole purpose of validating the control technique. Taking as a reference the work by McInnes in [7], here a similar Gaussian initial distribution profile is considered, and the target is defined in terms of uniform final density n_{ref} for all the shells. The feedback proportional control logic of Eq. (3), with the inputs in Table 1, acts on the deposition rate $\dot{n}_{launches}$, tuning the launch profile in time and space based on the local density of each region.

Table 1. Controller inputs of Eq. (3).

u_{Max}	e_{Max}	n_{ref}
$2e-7 \ [\#/sm^3]$	$0.5 \ [\#/m^3]$	$1 \ [\#/m^3]$

In Fig. 2a it is clearly visible the feedback effect in time to bring the density profile to the target one. The controlled launch rate is provided in Fig. 2b, and as expected the deposition rate is close to zero at high altitudes where the initial density is close to one and the absence of a relevant sink mechanism causes accumulation of objects. Differently, it is visible that with the constraints given in Table 1, the control action is not capable of overcoming the strong drag effect at low altitudes, even with the maximum allowed source rate, and after 100 days the profile reaches a limiting scenario. The results obtained agree with the analytically derived ones in [7].

This simple example paves the way to model more complex and realistic scenarios. All the contributions to Eq. (2) previously described will be modelled and many cases investigated, in terms of different inputs combinations, different sinks and sources profiles and different targets.

The proposed research responds to the need of a systematic way for analysing efficient strategies to face the debris proliferation problem. The approach is versatile, and the simplicity of this preliminary dynamical model allows for fast analyses. The tool will support the redefinition of regulations and standards for a sustainable utilisation of the space environment.

Acknowledgements
The research received funding from the European Research Council (ERC) under the European Union's Horizon Europe research and innovation program as part of the GREEN SPECIES project (Grant agreement No 101089265) and was partially supported by the ICSC—Centro Nazionale di Ricerca in High Performance Computing, Big Data, and Quantum Computing funded by European Union—NextGenerationEU.

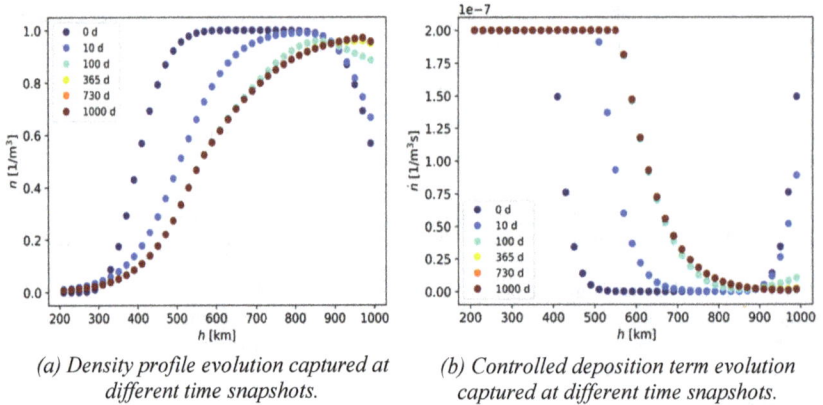

(a) Density profile evolution captured at different time snapshots.

(b) Controlled deposition term evolution captured at different time snapshots.

Fig. 2. Time evolution of the density and density rate profiles in altitude. Each point represents the value of one orbital shell. Snapshots are taken at 0, 10, 100, 365, 730 and 1000 days.

References

[1] ESA Space Debris Office. 2023. ESA' s Annual Space Environment Report, 2023.

[2] Inter-Agency Debris Committee. IADC Space Debris Mitigation Guidelines Issued by IADC Steering Group and Working Group 4, 2021.

[3] F. Letizia, B. Bastida Virgili, and S. Lemmens. Assessment of environmental capacity thresholds through long-term simulations. 72nd International Astronautical Congress, 2021. https://doi.org/10.1016/j.asr.2022.06.010

[4] ESA Space Debris Mitigation Working Group. ESA Space Debris Mitigation Requirements, Issue 1, 2023.

[5] H. G. Lewis, A. E. White, R. Crowther, and H. Stokes. Synergy of debris mitigation and removal. Acta Astronautica, vol. 81, no. 1, pp. 62–68, 2012. https://doi.org/10.1016/j.actaastro.2012.06.012

[6] R. Walker, C. E. Martin, P. H. Stokes, J. E. Wilkinson, and H. Klinkrad. Analysis of the effectiveness of space debris mitigation measures using the delta model. Advances in Space Research, vol. 28, no. 9, pp. 1437-1445, 2021. https://doi.org/10.1016/S0273-1177(01)00445-8

[7] C. R. McInnes. Simple analytic model of the long-term evolution of nanosatellite constellations. Journal of Guidance, Control, and Dynamics, vol. 23, no. 2, pp. 332–338, 2000. https://doi.org/10.2514/2.4527

[8] F. Letizia, C. Colombo, and H. G. Lewis. Analytical model for the propagation of small-debris-object clouds after fragmentations. Journal of Guidance, Control, and Dynamics, vol. 38, no. 8, pp. 1478–1491, 2015. https://doi.org/10.2514/1.G000695

[9] L. Giudici, C. Colombo, A. Horstmann, F. Letizia, and S. Lemmens. Density-based evolutionary model of the space debris environment in low-Earth orbit. Acta Astronautica, 2024. https://doi.org/10.1016/j.actaastro.2024.03.008

[10] S. Frey, C. Colombo, and S. Lemmens. Application of density-based propagation to fragment clouds using the Starling suite. 1st International Orbital Debris Conference, 2019.

[11] G. L. Somma, H. G. Lewis, and C. Colombo. Adaptive remediation of the space debris environment using feedback control. University of Southampton, Faculty of Engineering and Physical Sciences, PhD Thesis, 2019.

Aerospace Science and Engineering - IV Aerospace PhD-Days
Materials Research Proceedings 42 (2024) 117-120

Materials Research Forum LLC
https://doi.org/10.21741/9781644903193-26

Optimization strategies for a 16U CubeSat mission

Matteo Gemignani[1,a,b]

[1]Department of Civil and Industrial Engineering – Aerospace Division, University of Pisa, Via Girolamo Caruso 8, 56122, Pisa, PI, Italy

[a]matteo.gemignani@phd.unipi.it, [b]matteo.gemignani@unitn.it

Keywords: CubeSat, IOD/IOV, Green Propulsion, Electric Propulsion, Optimization, Small Satellites

Abstract. The surge in small satellite and CubeSat deployments has led to a diversification of feasible missions, driving a shift from emphasizing simplicity and low-cost to prioritizing performance while maintaining cost-efficiency. Integration of small payloads and advancements in technology have enhanced CubeSat capabilities, enabling the development of high-performance platforms. EXCITE, a 16U CubeSat mission, will demonstrate five different technologies in the LEO environment, including propulsion systems, a reconfigurable antenna, and on-board processing capabilities. To maximize EXCITE's capabilities and accommodate diverse payloads for various mission scenarios, multiple optimization strategies have been implemented. This includes thorough orbit analysis to determine the most suitable orbit for mission objectives, considering factors such as beta angles and eclipse time. The use of a chemical thruster provides flexibility in mission design by allowing adjustments to orbital altitude and ground-track patterns. Additionally, careful scheduling of orbital maneuvers is crucial for maximizing access time to specific ground stations and optimizing data downlink opportunities. Managing complex interactions between design variables necessitates advanced optimization techniques like gradient-based algorithms. OpenMDAO offers a robust framework for tackling multidisciplinary design optimization problems efficiently, facilitating exploration of trade-offs between competing design objectives.

Introduction

In recent years, there has been a remarkable surge in the deployment of small satellites and CubeSats into orbit annually. This proliferation has naturally expanded the variety of feasible missions achievable with these spacecrafts, encompassing in-orbit demonstrations, remote sensing, and scientific experimentation. Consequently, there has been a growing demand for improved performance across various aspects such as pointing accuracy, power generation, and data downloading [1]. This evolving demand has prompted a departure from the original attributes of CubeSat projects, which emphasized simplicity, low-cost, and high-risk, towards a new paradigm that prioritizes performance while still maintaining cost-efficiency.

Moreover, the compact nature of CubeSats, combined with advancements in technology, has enabled the integration of small payloads [2]. This addition further enhances the capabilities of CubeSats as IOD/IOV platforms, eventually equipped with propulsion systems which extend the possible mission design options [3]. Following this trend, a 16 U Cubesat mission named EXCITE has been developed by a collaborative effort led by the University of Pisa and four SMEs based in the Tuscany region: CRM Compositi, MBI, IngeniArs and Aerospazio Tecnologie. EXCITE will host 5 different technologies for Cubesat to be demonstrated in the LEO environment:

- *(CHIPS) Green monopropellant thruster*: a hydrogen peroxide propulsion system capable of performing moderate delta-V manoeuvres on a Cubesat, under development at UniPi.

- *(PPT) Pulsed plasma thruster*: a miniaturized electric thruster from Aerospazio Tecnologie S.r.l., capable of delivering very low impulse bits, for proximity operations or fine attitude control.
- *(REISAN) Reconfigurable integrated S-band antenna*: an electronically steerable antenna based on exciters distributed on suitable spacecraft surfaces, developed at UniPi.
- *(IoT-GPU) Internet-of-things GPU demodulator*: a technology by MBI S.r.l, based on the utilization of a COTS GPU for on-board processing of advanced IoT waveforms and VDES protocol.
- *(PHP) Pulsating heat pipes*: high throughput heat pipes under development at UniPi, based on unsteady fluid flow, especially suited for high heat flux applications (e.g., thermal management of high-power microsatellites).

Optimization Strategies

Since its early development, our focus has been on transforming EXCITE into a high-performance platform, driving us to implement multiple optimization strategies. Our goal is to maximize its capabilities and accommodate various payloads for diverse mission scenarios.

To address this process, we conducted a thorough analysis of potential orbits for the mission. Given its dimensions, EXCITE is likely to be hosted as a secondary payload on a VEGA-C launcher and placed into a Sun-Synchronous Orbit at approximately 550 km altitude. Such orbits are ideal for Earth Observation and provide global coverage [4].

However, the nature of Sun-Synchronous Orbits introduces considerations regarding beta angles, which significantly impact power generation and eclipse time. There are two extreme cases in this sense: dawn-dusk orbits, where the satellite traverses the termination line, minimizing eclipse time, and noon-midnight orbits, where eclipse time is maximal. Each option has its advantages and drawbacks. While no eclipse maximizes power production from solar panels, it may pose thermal management challenges. Conversely, a longer eclipse creates a dynamic thermal environment and reduces power production per orbit. Anyway, most SSO satellites for Earth observation are placed into orbits with a local time of ascending node (LTAN) around 10 A.M or 2 P.M [5]. This is optimal for having not too long shadows and good illumination conditions.

Considering EXCITE's role as a secondary payload, our spacecraft design must account for all possible Right Ascension of the Ascending Node (RAAN) values in which we could be deployed, which in turn will affect the eclipse time, as shown in Figure 1.

The possibility for EXCITE to use a chemical thruster gives us lots of possibilities regarding the mission design. In particular, by using the thruster to lower or elevate the orbital altitude it is possible to obtain different ground-track patterns. Figure 2 shows how orbits with periods which are dividers of a daytime, have repeated ground track patterns, and are suitable for example to fly over specific targets every day at the same hour, while different altitudes provide better coverage over a region for interest, for example the North of Italy. This proves how a propulsion system on a Cubesat improves the mission design flexibility and also provides adjustability to any orbit insertion error after the deployment of the satellite [6]. For this reason, the amount of propellant and the orbital maneuvers schedule will be crucial design variables for the optimization of EXCITE.

Furthermore, the scheduling of orbital maneuvers is essential for maximizing access time on specific ground stations. By carefully planning and executing maneuvers, the CubeSat can align its ground track with the desired ground station passes, thereby maximizing communication windows and data downlink opportunities.

Materials Research Forum LLC

https://doi.org/10.21741/9781644903193-26

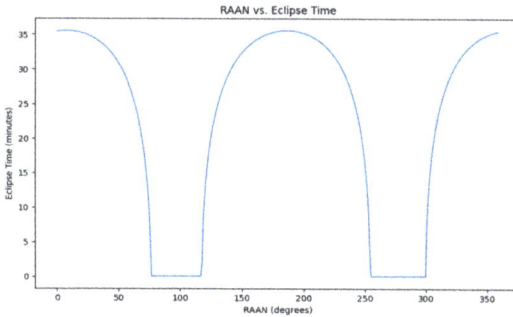

Figure 1: Eclipse time vs RAAN for a Sun-Synchronous Orbit at 550 km altitude

For example, using the CHIPS propulsion system, assuming 1 kg of propellant mass, it will be possible to perform a Hohmann transfer from 550 km altitude to 500 km, using about 40% of total propellant mass. Then the rest of the propellant might be used to compensate the drag losses, extending the mission duration, or for a faster de-orbit at the end of the spacecraft lifetime.

Thermal management is another critical aspect of mission optimization, ensuring that onboard systems operate within their specified temperature ranges for optimal performance. This is nowadays of great interest among small satellite manufacturers because, with higher available power on-board, thermal management for highly integrated and compact platforms has become difficult. Pulsating Heat Pipes (PHP) installed on-board of EXCITE will dispose of the heat building up on the body-mounted solar panel. Compared with other two-phase passive thermal devices, such as conventional heat pipes and loop heat pipes , PHPs have many advantages, including having a simple construction, being lightweight and flexible, and having no internal wick structure. Moreover, they can be embedded inside the composites structure of the Cubesat. For this reason, their layout is also a factor to be considered in optimization.

Power production, orbit changes schedule, thermal management, and the design of the composites structure of EXCITE will have a huge impact on the attitude control system, both on the design and operations schedule.

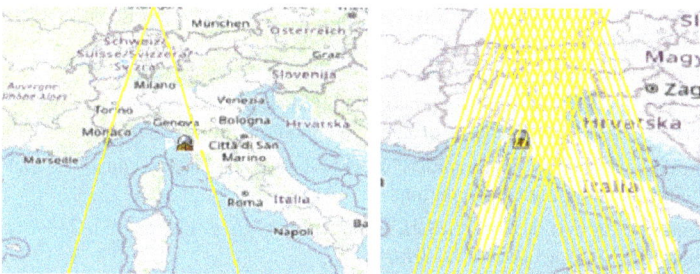

Figure 2: Ground Tracks over Italy propagated for 15 days at altitudes of 555 km (left) and 570 km (right). The tracks on the left image overlap.

Finding an optimal design for a system with so many highly coupled design variables result in a non-trivial mathematical problem. Managing such a high-dimensional optimization problem effectively requires advanced optimization techniques capable of handling large-scale, multidisciplinary design optimization (MDO) problems. Due to this fact, we are considering gradient-based algorithms, such as those provided by OpenMDAO (Open-source

Multidisciplinary Design Analysis and Optimization) [7] that offers a robust framework for tackling such optimization challenges. By leveraging gradients or derivatives of the objective function and constraints with respect to the design variables, gradient-based algorithms can efficiently navigate the design space to identify optimal solutions. This approach enables faster convergence to high-quality solutions and facilitates the exploration of trade-offs between competing design objectives.

Conclusions

Due to the diversity that EXCITE, and IOD Cubesats in general can accommodate, thorough optimization strategies, encompassing orbit analysis, propulsion system utilization, and thermal management, are crucial for maximizing mission effectiveness. Advanced optimization techniques, such as gradient-based algorithms offered by OpenMDAO, play a pivotal role in navigating the complex interplay of design variables and achieving optimal solutions. Ultimately, EXCITE serves as a testament to the integration of cutting-edge technologies and optimization methodologies, ensuring efficient and effective mission outcomes in the dynamic realm of small satellites design.

Acknowledgments

This extended abstract was produced while attending the PhD program in PhD in Space Science and Technology at the University of Trento, Cycle XXXIX, with the support of a scholarship financed by the Ministerial Decree no. 118 of 2nd march 2023, based on the NRRP - funded by the European Union - NextGenerationEU - Mission 4 "Education and Research", Component 1 "Enhancement of the offer of educational services: from nurseries to universities" - Investment 4.1 "Extension of the number of research doctorates and innovative doctorates for public administration and cultural heritage".

References

[1] N. Saeed, A. Elzanaty, H. Almorad, H. Dahrouj, T. Y. Al-Naffouri, and M.-S. Alouini, "CubeSat Communications: Recent Advances and Future Challenges," Aug. 2019, [Online]. Available: http://arxiv.org/abs/1908.09501

[2] M. Mozzato et al., "Concept and Feasibility Analysis of the Alba Cubesat Mission," Aerotecnica Missili & Spazio, Mar. 2024. https://doi.org/10.1007/s42496-024-00205-9

[3] A. Frolova, L. Kuhlmann, and J. Silveira, "Sustainability Aspects of Rapid Prototyping and Frequent In-Orbit Demonstrations with CubeSats," 2023. https://doi.org/10.13009/EUCASS2023-691

[4] R. J. Boain, "A-B-Cs of Sun-Synchronous Orbit Mission Design 14 AAS/AIAA Space Flight Mechanics Conference A-B-Cs of Sun-Synchronous Orbit Mission Design."

[5] Arianespace, "All flight opportunities." Accessed: Mar. 13, 2024. [Online]. Available: https://smallsats.arianespace.com/opportunities

[6] K. Lemmer, "Propulsion for CubeSats," Acta Astronaut, vol. 134, pp. 231–243, May 2017. https://doi.org/10.1016/j.actaastro.2017.01.048

[7] J. S. Gray, J. T. Hwang, J. R. R. A. Martins, K. T. Moore, and B. A. Naylor, "OpenMDAO: an open-source framework for multidisciplinary design, analysis, and optimization," Structural and Multidisciplinary Optimization, vol. 59, no. 4, pp. 1075–1104, Apr. 2019. https://doi.org/10.1007/s00158-019-02211-z

Aerospace Science and Engineering - IV Aerospace PhD-Days   Materials Research Forum LLC
Materials Research Proceedings 42 (2024) 121-125   https://doi.org/10.21741/9781644903193-27

Wing-aileron adaptive flutter suppression system

Carmelo Rosario Vindigni[1,a] *, Calogero Orlando[1,b], Antonio Esposito[1,c] and Andrea Alaimo[1,d]

[1]Kore University of Enna, M.A.R.T.A. Centre, Enna, 94100, Italy

[a]carmelorosario.vindigni@unikore.it, [b]calogero.orlando@unikore.it, [c]antonio.esposito@unikore.it, [d]andrea.alaimo@unikore.it

Keywords: Wing Aileron Stick Model, Flutter Suppression, Simple Adaptive Control

Abstract. In this work a flutter suppression system design based on simple adaptive control architecture and an alternative beam finite element modelling of wings equipped with trailing edge control surfaces is proposed. The aeroelastic beam finite element used is based on Euler-Bernoulli beam theory for the flexural behavior, De Saint Venant theory for torsion and two-dimensional time-domain unsteady aerodynamics applied by means of strip theory assumptions. The finite element modeling used allows to write the aero-servo-elastic plant governing equations in state-space form, from which the flutter suppression system design can be carried out in a time domain fashion. The simple adaptive control architecture has been applied to the aero-servo-elastic plant which passivity requirement has been enforced implementing a parallel feed-forward compensator.

Introduction

Over the years, wing structures design has been influenced by the preference of realizing lightweight structural configurations; however, this choice have led to face the problem of structural susceptibility to aeroelastic instabilities [1]. An open research field deals with active suppression approaches that are related to control systems implementation and actuation methods in order to stave off the instabilities by suppressing the related vibration. Flutter suppression systems design based on equivalent beam modelling of the aeroelastic plant have been studied in literature following different approaches, but the implementation of Simple Adaptive Control SAC architecture in this framework has not been explored yet. In fact, the SAC scheme has been successfully applied in literature for the flutter suppression of three degrees of freedom 3DOF airfoil models; in detail, in [2] an Implicit Reference Model IRM based SAC architecture has been developed, in [3] a SAC algorithm is used to suppress the flutter vibration of a 3DOF airfoil with pitch stiffness non-linearity, where the trailing edge control surface is deployed by means of a piezoelectric actuator, increasing the flutter boundary of about the 213% with respect to the open loop case, and the robustness of this latter aero-servo-elastic plant is studied for structural, aerodynamics, actuator, and free-play uncertainties. However, the 3DOF airfoil is a low fidelity model not representative of wings with locally distributed control surfaces; therefore, this work aims to assess the performances of the SAC flutter suppression scheme when applied to a more realistic aeroelastic plant such as the one obtained implementing an equivalent beam system.

Problem statement and finite element formulation

The aeroelastic system considered in this work is a cantilever wing with semi-span length l_w, straight elastic axis EA, and a trailing edge aileron-like control surface that extends from a distance l_F by the root to the wing tip. The aileron control surface is hinged to the wing frame; the links between the servo and the aileron hinge are also taken into account and their positions with respect to the l_F wing station are defined as l_{a1} and l_{a2}. The control surface is flexible in torsion and with elastic axis close enough to the hinge line such that they could be considered coincident. The wing is modelled as an equivalent beam according to Euler-Bernoulli and De Saint Venant beam theories assumptions. The structural degrees of freedom of the wing-stick model are the vertical displacement due to bending $w(x, t)$ (positive upwards), the elastic torsional rotation $\phi\,(x, t)$ around the elastic axis (positive nose-up), and the control surface rotation $\delta\,(x, t)$ about its hinge (positive aileron down). The wing's parameters are shown in Figure 1. In detail, b is the semi-chord, a is the non-dimensional distance from the mid-chord to the elastic axis (positive aft), x_ϕ is the non-dimensional distance between the mid-chord and the center of gravity (positive aft), c identifies the control surface hinge location behind the mid-chord, while its center of gravity lies at x_δ behind the hinge.

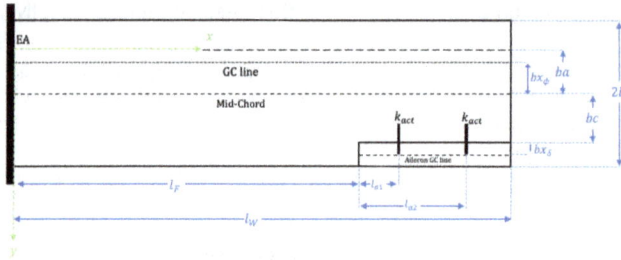

FIGURE 1: Wing-aileron schematic

The wing governing equations are the ones of a beam with bending and torsion degrees of freedom to which is added the torsional equation of motion of the aileron. A compact matrix representation of the governing equations can be obtained properly defining the structural and aerodynamic matrices that represent mass $\mathbf{M}_s$, $\mathbf{M}_{aer}$, damping $\mathbf{C}_s$, $\mathbf{C}_{aer}$, and stiffness $\mathbf{K}_s$, $\mathbf{K}_{aer}$ contributions, where the subscripts s and aer stand for structural and aerodynamic, respectively. Moreover, the gust loads contribution is introduced as $\mathbf{F}^g w_g(t)$. Thus, defining a generalized displacement vector $\mathbf{q}(t) = [w \quad \phi \quad \delta \quad \bar{x}]^T$ the governing equations compact form reads as

$$[\mathbf{M}_s + \mathbf{M}_{aer}]\ddot{\mathbf{q}}(t) + [\mathbf{C}_s + \mathbf{C}_{aer}]\dot{\mathbf{q}}(t) + [\mathbf{K}_s D^2 + \mathbf{K}_\delta + \mathbf{K}_{aer}]\mathbf{q}(t) = \mathbf{F}_g w_g(t) \tag{1}$$

where D is a differential operator related to the derivatives along the beam axis and $\mathbf{K}_\delta$ is the aileron stiffness matrix that collects the hinge and actuator stiffnesses. The matrices full expressions can be found in [4]. Introducing the displacement field interpolation and writing eq. 1 in weak form the aeroelastic beam elemental mass, damping and stiffness matrices are obtained performing the following integrals

$$\mathbf{M}_e = \int_L \mathbf{N}^T [\mathbf{M}_s + \mathbf{M}_{aer}]\mathbf{N}dx; \; \mathbf{C}_e = \int_L \mathbf{N}^T [\mathbf{C}_s + \mathbf{C}_{aer}]\mathbf{N}dx; \; \mathbf{K}_e = \int_L [(D\mathbf{N})^T \mathbf{K}_s D + \mathbf{N}^T \mathbf{K}_\delta + \mathbf{N}^T \mathbf{K}_{aer}]\mathbf{N}dx \tag{2}$$

Then, assembling the matrices opportunely, in accordance with the wing discretization, and imposing the boundary conditions, the structural equations of motion reads as

$$\begin{bmatrix} \mathbf{M}_{11} & \mathbf{M}_{12} \\ \mathbf{M}_{12}{}^T & \mathbf{M}_{22} \end{bmatrix} \begin{bmatrix} \ddot{\Delta}_1 \\ \ddot{\Delta}_2 \end{bmatrix} + \begin{bmatrix} \mathbf{C}_{11} & \mathbf{C}_{12} \\ \mathbf{C}_{12}{}^T & \mathbf{C}_{22} \end{bmatrix} \begin{bmatrix} \dot{\Delta}_1 \\ \dot{\Delta}_2 \end{bmatrix} + \begin{bmatrix} \mathbf{K}_{11} & \mathbf{K}_{12} \\ \mathbf{K}_{12}{}^T & \mathbf{K}_{22} \end{bmatrix} \begin{bmatrix} \Delta_1 \\ \Delta_2 \end{bmatrix} = \begin{bmatrix} \mathbf{F}_S^g \\ 0 \end{bmatrix} w_g \tag{3}$$

where Δ_1 and Δ_2 are the unknown and known displacements vectors, from which the rearrangement of the structure matrices is performed, while $\mathbf{F}_S^g w_g$ is the global gust force vector. Then, eq. 3 can be cast in state space form introducing the state vector $X = [\Delta_1^T \quad \dot{\Delta}_1^T]^T$, that collects the unknown structural displacements and their time derivative, and the dynamic matrix computed as follows

$$A = \begin{bmatrix} 0 & I \\ -M_G^{-1}K_G & -M_G^{-1}C_G \end{bmatrix} \tag{4}$$

Moreover, in order to obtain the state-input relation **B**, a submatrix K_{12}^{δ} of K_{12}, corresponding to the aileron displacement at the actuator-aileron linking stations, must be identified. The **B** matrix is computed considering the aileron stiffness only since the related mass and damping contributions have shown to be negligible for the aileron dynamics in correspondence of the actuators. Thus, the state-input matrix and the state input relation for the gust loads read as

$$B = \begin{bmatrix} 0 \\ -M_{11}^{-1}K^{\delta}_{12} \end{bmatrix} ; B^g = \begin{bmatrix} 0 \\ -M_{11}^{-1}F_S^g \end{bmatrix} \tag{5}$$

Last, the state-output matrix **C** is an identity matrix in the hypothesis of ideal sensors; therefore, the wing state space system read as

$$\begin{cases} \dot{X} = AX + Bu + B^g w_g \\ \quad Y = CX \end{cases} \tag{6}$$

where u is the control input, i.e. the aileron displacement at the actuators stations, and w_g is the gust profile.

Adaptive controller
The SAC, that could be defined as a modification of the Model Reference Adaptive Control, takes into account a reference model that generates the signal to be tracked by the controlled plant in order to let it follow the desired dynamics [5].In detail, the SAC control signal is a linear combination of the reference model state, input, and of the tracking error $e(t)$; however,in this work the control objective is to make the wing tip torsion angle zeroed, thus the SAC control law is reduced to [6] taking into account the output signal $e(t) = 0 - \phi(l_w, t)$, only. Thus, the output feedback adaptive control law reads as

$$u(t) = (K_{eP} + K_{eI})e(t) \tag{7}$$

where K_{eP} and K_{eI} are the controller adaptive gains expressed as $K_{eP} = \Gamma_{eP}e^2$ and $\dot{K}_{eI} + \eta K_{eI} = \Gamma_{eI}e^2$, with Γ_{eP} and Γ_{eI} invariant gains of the control algorithm and η the Iannou term that ensures the stability of the system when bounded disturbances are present [6]. However, it is to be said that the SAC algorithm can be only applied to systems that fulfill the passivity requirements [7]. In this work, a Parallel Feedforward Compensator PFC is added to the plant $G(s)$, i.e the Single Input Single Output SISO transfer function related to Eq. 6, in order to make the system meet the Almost Strictly Positive Realness ASPR condition [8]. As presented in [9], the PFC is designed as the inverse of a controller $H(s)$ that stabilize the closed loop system given by $G_c(s) = (1 + G(s)H(s))^{-1}G(s)$. The selected controller $H(s)$ is a ideal Proportional-Derivative PD controller with transfer function $H(s) = K_H(1 + \tau_H s)$, being $K_H = 3$ and $\tau_H = 10^{-3}s$, that makes the augmented plant satisfy the ASPR condition for every speed values below the SAC closed loop system flutter boundary $v_f^{SAC} = 171\ m/s$. A scheme of the SAC control architecture specialized for the wing-aileron flutter suppression is shown in Figure 2.

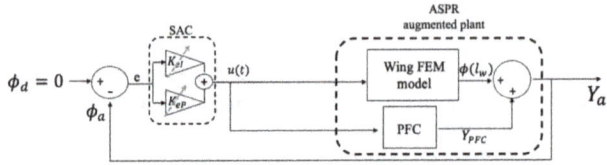

FIGURE 2: Wing-aileron SAC flutter suppression architecture

The invariant gains of the adaptive controller, namely $[\Gamma_{eP} \quad \Gamma_{eI} \quad \eta]$, are tuned using a Population Decline Swarm Optimization P_DSO algorithm [10] and considering the plant at the open loop flutter speed $v_f^{OL} = 109.5 \, m/s$ subjected to a pulse disturbance on the tip torsion rate with amplitude $\dot{\phi}(l_w, 0) = 100 \, rad/s$ occurring at time instant $t = 0$ and with width $t = 0.001s$. From the results of the optimization, the invariant gains are selected as $[3.046 \times 10^6 \quad 304 \quad 10^{-3}]$ providing for the minimum objective function value $ITAE_{min} = 3.71 \times 10^{-5} \, rad * s$.

Closed loop system analysis

Numerical simulations are carried out to study the performances of the SAC flutter suppression system in flight scenarios of interest where the wing is perturbed by discrete atmospheric gusts. In detail, in order to be confident with real flight scenarios, a 1−cosine discrete gust profile has been considered, as suggested by references [11,12]. One case study considered involves the wing flying at the flutter speed and simultaneously encountering a pulse disturbance on the tip torsion rate and a 1−cosine gust with maximum peak $w_{max} = 1 \, m/s$ and gust semi-width $t_g = 0.25$. The simulation results are reported in Figure 3 where it can be observed that the flutter oscillations are suppressed in less that 2 seconds and kinematic variables peak values are reduced, with respect to the open loop case, with a maximum aileron deflection of 0.8deg. In detail, the tip torsion is reduced from −0.54deg to −0.43deg and the tip deflection from 0.037m to 0.028m.

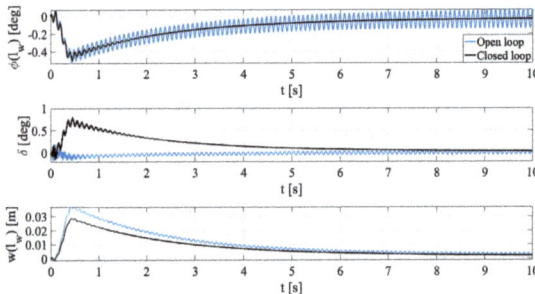

FIGURE 3: System response at the flutter speed with gust disturbance

Conclusions

This work has presented an alternative aeroelastic beam finite element method for the numerical modeling of wing structures with aileron-like control surfaces. The numerical model generated offers a suitable state space description of the aeroelastic system and has been tailored for flutter suppression system design objectives. The Adaptive Controller's invariant gains have been tuned using a meta-heuristic swarm technique called P_DSO. The adaptive flutter suppression system designed has demonstrated the ability to increase the wing flutter boundary by the 55.25%.

Aerospace Science and Engineering - III Aerospace PhD-Days   Materials Research Forum LLC
Materials Research Proceedings 42 (2024) 121-125   https://doi.org/10.21741/9781644903193-27

Numerical simulations in the presence of gust disturbance have been used to examine the performance of the closed loop system that have shown a satisfactory dynamic behavior.

Funding

The study was financially supported by the Italian Ministry of University and Research - M.U.R. under the DAVYD project (P.O.N. Grant ARS01_00940).

References

[1] E. Livne, Aircraft active flutter suppression: State of the art and technology maturation needs. *Journal of Aircraft*, 55(1), 410-452, 2018.

[2] B. Andrievsky, E. V. Kudryashova, N. V. Kuznetsov, O. A. Kuznetsova, and G. A. Leonov, "Simple adaptive control for airfoil flutter suppression," Mathematics in Engineering, Science and Aerospace, vol. 9, no. 1, pp. 5–20, 2018.

[3] C. R. Vindigni, A. Esposito, and C. Orlando, "Stochastic aeroservoelastic analysis of a flapped airfoil," Aerospace Science and Technology, vol. 131, p. 107967, 2022.

[4] C. R. Vindigni, G. Mantegna, A. Esposito, C. Orlando, and A. Alaimo, "An aeroelastic beam finite element for time domain preliminary aeroelastic analysis," Mechanics of Advanced Materials and Structures, pp. 1–9, 2022.

[5] I. Barkana, "Simple adaptive control–a stable direct model reference adaptive control methodology–brief survey," International Journal of Adaptive Control and Signal Processing, vol. 28, no. 7-8, pp. 567–603, 2014.

[6] A. L. Fradkov, I. V. Miroshnik, and V. O. Nikiforov, Nonlinear and adaptive control of complex systems, vol. 491. Springer Science & Business Media, 2013.

[7] I. Barkana, "Robustness and perfect tracking in simple adaptive control," International Journal of Adaptive Control and Signal Processing, vol. 30, no. 8-10, pp. 1118–1151, 2016.

[8] I. Rusnak and I. Barkana, "Spr and aspr untangled," IFAC Proceedings Volumes, vol. 42, no. 6, pp. 126–131, 2009.

[9] C. R., Vindigni, G. Mantegna, C. Orlando, and A. Alaimo, Simple adaptive wing-aileron flutter suppression system. *Journal of Sound and Vibration*, 570, 118151, 2024.

[10] Vindigni, C. R., & Orlando, C. (2020). A gain scheduling control of incompressible airfoil flutter tuned by the population decline swarm optimizer—PDSO. Aerotecnica Missili & Spazio, 99(1), 3-16.

[11] R. L. Bisplinghoff and H. Ashley, Principles of aeroelasticity. Courier Corporation, 2013.

[12] Dowell, Earl H., *A modern course in aeroelasticity*. Vol. 264. Springer Nature, 2021.

Aerospace Science and Engineering - IV Aerospace PhD-Days
Materials Research Proceedings 42 (2024) 126-131

Materials Research Forum LLC
https://doi.org/10.21741/9781644903193-28

Multi-fidelity approach for sonic boom annoyance

Samuele Graziani[1,a] *, Nicole Viola[1,b] and Roberta Fusaro[1,c]

[1]Corso Duca degli Abruzzi 24, Turin, 10129, Italy

[a]samuele.graziani@polito.it, [b]nicole.viola@polito.it, [c]roberta.fusaro@polito.it

Keywords: Sonic Boom, CFD, Aeroacoustics, Psychoacoustics

Abstract. The following abstract aims at summarizing the multi-fidelity approach developed in the framework of the H2020 MORE&LESS project for the evaluation of sonic boom phenomena from a physical and psychoacoustic point of views for different aircraft configurations. Starting from the conceptual design phase, new analytical formulations shall be developed and validated to accurately define sonic boom without the necessity of having high time-consuming simulations. Then, once a 3D CAD model is available, numerical higher-fidelity simulations can be carried out. This step consists of both the near-field CFD and the use of a propagation code to propagate the shocks from the aircraft altitude to the ground. The H2020 MORE&LESS offers the opportunity to validate formulations and numerical results thanks to open-field test campaigns with small-scale aircraft models. Finally, having all information regarding the ground signature, a psychoacoustics code can be employed to define the annoyance caused by sonic boom comparing different noise metrics. Throughout the paper, the preliminary results available for a Mach 5 waverider configuration are reported.

Introduction

Over the past few decades, several initiatives have aimed to create a civil supersonic aircraft capable of succeeding the Concorde [1]

However, these endeavors have encountered significant hurdles, ranging from environmental concerns to economic viability, efficiency, and the challenge of mitigating the effects of sonic booms. One of the primary obstacles facing the development of the next generation of supersonic aircraft is the need to reduce the impact of sonic booms to levels acceptable to the general population. Since 1973, supersonic flight over land for civil aircraft has been prohibited due to the disturbance caused by sonic booms. Despite the absence of clear regulations for supersonic flight over land, NASA outlined objectives in 2011 for future supersonic aircraft, which included criteria for acceptable sonic boom levels. [2] In recent years, considerable advancements have been made in the start-of-the-art technologies aimed at minimizing sonic boom and ensuring criteria. The new generation of supersonic transport (SST) aims to establish standards of acceptability for supersonic flight over land during cruising phases and for noise levels during takeoff and landing operations near airports. [3] However, the most influential factors affecting the sonic boom are linked to the aircraft's shape, as well as flight parameters such as Mach number, altitude, weight, and flight path angle. In this context, the H2020 MORE&LESS project has been funded by the European Commission to develop a multi-fidelity approach that can swiftly and accurately predict the sonic boom signature of new aircraft designs from the conceptual stage becomes paramount. Nevertheless, during this phase of the design process, conducting sensitivity analyses is essential for thoroughly evaluating the sonic boom throughout the mission profile.

Aerospace Science and Engineering - III Aerospace PhD-Days   Materials Research Forum LLC
Materials Research Proceedings 42 (2024) 126-131   https://doi.org/10.21741/9781644903193-28

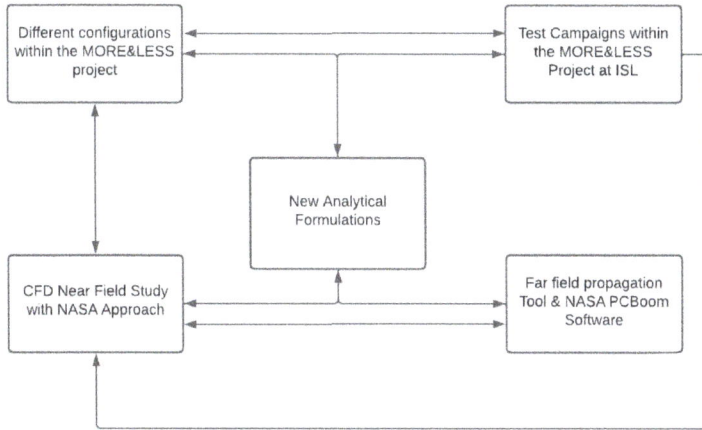

Figure 1: Scheme for New Analytical Formulations

Case study

In this framework, a mockup configuration used during a test campaign of one of the case studies of the MORE&LESS project is used. The experimental tests were carried out at ISL's facility in France. The original aircraft was re-scaled and suitably modified to meet the aerodynamic and structural requirements following the sabot separation. The final configuration possesses a mass of just over 500 grams, a length of 20.16 cm, with the center of gravity positioned approximately 11.2 cm from the aircraft's nose. The stability was evaluated both via CFD simulations and wind-tunnel tests. The results provided via CFD, and propagation tool are compared to the acoustic field microphones that are a Bruel & Kjaer 4938 ¼" microphones placed on the ground at different distances from the centerline. The results highlight a different behavior depending on roll angle region, with a global behavior in which the mean value increases with roll angle value.

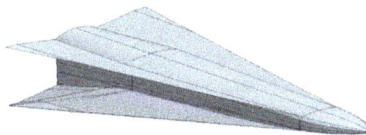

Figure 2: Model Used in the test campaign

Analytical formulations to evaluate sonic boom

There are many analytical models documented in literature that evaluate sonic booms during conceptual design phases. Among these, Carlson's simplified method [4] stands out as particularly renowned. This method offers a procedure for computing N-Wave sonic booms during steady flight or moderate descent/climb in a standard atmosphere. Its simplification lies in the reformulation of the Whitham F-function, which is approximated to a constant known as the aircraft shape factor. This factor varies based on the geometric and flight conditions of the aircraft. The first step in evaluating the sonic boom of a supersonic configuration using this methodology is the estimation of the equivalent area due to volume, which is the cross-sectional area normal to the Mach cone. The following step is to define the equivalent area due to lift, which is approximated with the planform area distribution.

The final step consists in the combination of the two previous contributions to obtain the total effective area of the aircraft. With so, the evaluation of bow shock overpressure is written as:

$$\Delta p = K_p \cdot K_r \cdot \sqrt{p_v \cdot p_g} \cdot (M^2 - 1)^{\frac{1}{8}} \cdot h_e^{-\frac{3}{4}} \cdot l^{\frac{3}{4}} \cdot K_s \qquad (1)$$

The time signature duration can be evaluated as:

$$\Delta t = K_t \cdot \frac{3.42}{a_v} \cdot \frac{M}{(M^2-1)^{\frac{1}{8}}} \cdot h_e^{\frac{1}{4}} \cdot l^{\frac{3}{4}} \cdot K_s \qquad (2)$$

Having the results of the experimental tests, validated with the numerical high-fidelity simulations, is possible to modify and improve the analytical formulations for better describing different configurations of supersonic and hypersonic aircraft.

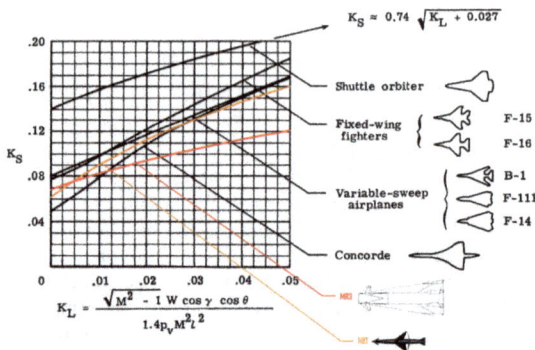

Figure 3:Modification of the Formulation for different test cases

Numerical high-fidelity simulations

Sonic boom prediction involves several complex physical phenomena, aerodynamics instabilities, pressure disturbances and high-speed aerodynamic flows. Numerical simulations provide an accurate estimation of the near-field aerodynamic perturbation. To accurately define the fluid domain and the nonlinearities for capturing the proper sonic boom signature, it is essential to have a proper numerical discretization and precision of advective terms.

Since 2014, NASA has been promoting the Sonic Boom Prediction Workshop [5-6-7], aimed at refining numerical methods for predicting sonic booms. The workshop's findings have paved the way for establishing guidelines on computationally forecasting noise generated by supersonic aircraft. Specifically, attention is focused on mesh strategies, including refinement and adaptation approaches, to accurately track discontinuities, demanding specific considerations. The computational grids topology adopts a hybrid methodology, comprising two distinct parts: an unstructured region close to the aircraft geometry and a structured segment for the distant domain. The unstructured portion resembles a cylinder integrated into the domain, while the structured section is constructed utilizing a blocking technique. The numerical elements are hexahedral in the structured part, tetrahedral in the unstructured zone, and pyramids to connect the two zones.

Aerospace Science and Engineering - III Aerospace PhD-Days   Materials Research Forum LLC
Materials Research Proceedings 42 (2024) 126-131   https://doi.org/10.21741/9781644903193-28

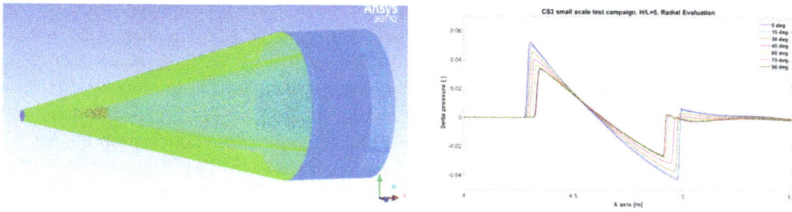

Figure 4: CFD Mesh and Signature H/L=5

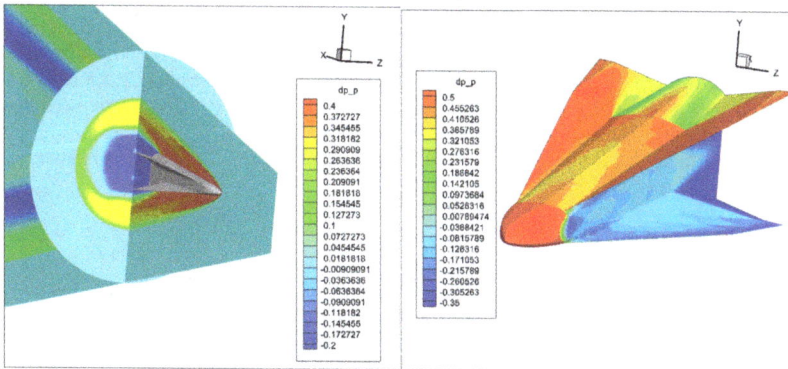

Figure 5: Contour of Delta Pressure distribution

To simulate the propagation of sonic boom shock waves reaching the ground, there is the need to employ a detailed far-field propagation algorithm. Unlike Computational Fluid Dynamics (CFD) simulations that focus on analyzing the pressure field around the aircraft, our tool prioritizes accounting for atmospheric inhomogeneities while neglecting wave refraction effects. It operates based on the Augmented Burgers Equation specifically formulated for sonic boom propagation, comprising a 1-D nonlinear wave equation assessed within the unidimensional realm of the acoustic ray.

Prior to simulation, acoustic rays must be determined using linear principles. The software utilizes the Augmented Burgers Equation to propagate the sonic boom shock waves along these acoustic rays, incorporating a stratified atmospheric profile encompassing pressure, wind, temperature, and relative humidity. Then it was able to derive pressure signature lines from CFD simulations conducted to evaluate local pressure near the aircraft. These signatures are extracted along lines parallel to the vehicle's trajectory, their positioning defined by azimuth angle ϕ and radial distance H. In preparation for propagation, a slight fading out is introduced at the end of the pressure signatures to ensure a smooth return to zero. Additional zeros, known as zero-padding, are inserted before propagation to optimize the computational process.

Annoyance metrics

The selection of proper noise metric for predicting human response to the low-amplitude sonic booms is the last key focus in the evaluation of the future impact of civil supersonic flights overland. The metrics that are currently under investigations include both engineering metrics, loudness metrics that considers the human perception of sounds and hybrid metrics that combines several metrics into one model.

An enroute Noise standard is the biggest challenge in supersonic aviation due to the requirements in new design approaches and due to the lack of relevant data to define limits.

Perceived Level has been widely used to describe the loudness of sonic boom and it is used as a target when optimizing supersonic aircraft and it works well in the explanation of human annoyance to outdoor booms. However, since man spend more than 70% of their time indoor, there is the need to define proper indoor metrics. Several alternate metrics have been proposed that treat lower frequencies differently, which is the critical point for the description of sonic boom noise.

In collaboration with University of Florida, a routine that will utilize prediction from the MOST program to analyze Sonic Boom metrics such as PL, ASEL, BSEL, DSEL, ESEL and ISBAP that employ post-processing routine to transform predicted sonic boom waveforms into annoyance metrics, thereby informing the design process.

In this paper, the steps for the evaluation of the Perceived Noise Level are highlighted using the Stevens Mark VII method [9]. In the method, the standard reference sound is defined as a 1/3 octave band centered at 3150 Hz. Starting from the band spectrum of noise, the procedure manages to predict the level of a reference signal that will be judged equal to the noise. Each band level is converted into a perceived value in sones, and the total perceived value is computed with a summation rule and converted in perceived level in dB called PLdB.

Having evaluate the sum of the perceived value for every sones, the final formulation for the calculation of the perceived level is:

$$P = 32 + 9 \cdot \log_2 S_t \tag{3}$$

Two plots of the procedure for the CS3 are viewed in Figure 6. In particular, the first one highlights the Sound Pressure Level (SPL) as a function of frequency for the various bands and the second one is Sonic Boom signature in the frequency domain. Since the extraction was made at H/L=3, the response is at high frequency compared to a classical signature on the ground.

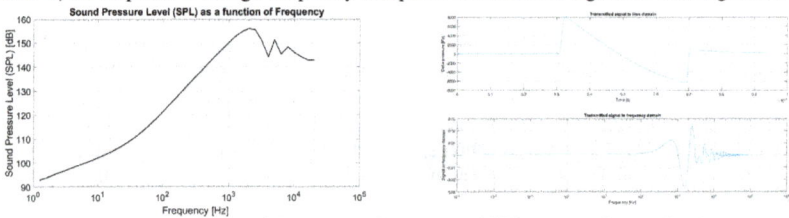

Figure 6: SPL and Frequency Spectrum of CS3 case study at H/L=3

The script is validated within a NASA test case for a low boom configuration as could be seen in Figure 7. However, in the NASA code, the application of a filter around 1000 Hz frequency is evident.

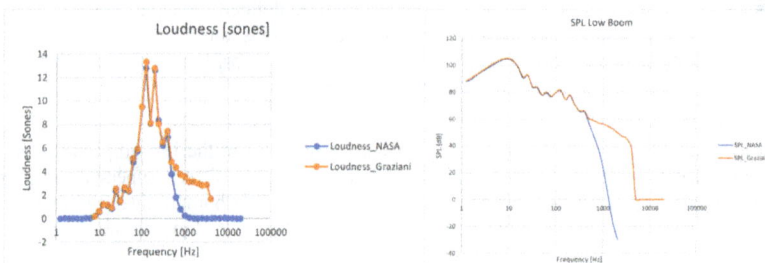

Figure 7: Psychoacoustic code validation with respect to NASA results

Aerospace Science and Engineering - III Aerospace PhD-Days   Materials Research Forum LLC
Materials Research Proceedings 42 (2024) 126-131   https://doi.org/10.21741/9781644903193-28

Conclusions

In the following paper, two methods for studying sonic booms are presented. The first consists of a low-fidelity analytical methodology adaptable to conceptual design, which does not require a high computational cost and does not require a large amount of data.

The second methodology consists of a high-fidelity study consisting of CFD for the near-field part near the aircraft, a propagation code based on the Augmented Burger Equation. Finally, for the second methodology, a psychoacoustic code was applied to assess the Perceived Noise Level according to Stevens' Mark VII method. The code was validated against an internal case study proposed by NASA.

The high-fidelity methodology could be replicated for any type of supersonic configuration, and a sensitivity analysis could be carried out to investigate the climate effects (such as humidity, temperature, winds) in the psychoacoustic response.

References

[1] B. Liebhardt, K. Lütjens, An Analysis of the market environment for supersonic business jets, 2011

[2] Advanced Concept Studies for Supersonic Commercial Transport Aircraft Entering Service in the 2030-2035 Period, 2011.

[3] Y. Sun, H. Smith, Review and prospect of supersonic business jet design. Progress in Aerospace Science, 2017, Volume 90, pages 12-38. https://doi.org/10.1016/j.paerosci.2016.12.003

[4] H. Carlson, Simplified Sonic-Boom Prediction, NASA technical paper 1122, 1978

[5] M. Park, J. Morgenstern, Summary and Statistical Analysis of the First AIAA Sonic Boom Prediction Workshop, 2014. https://doi.org/10.2514/6.2014-2006

[6] M. Park, M. Nemec, Nearfield Summary and Statistical Analysis of the Second AIAA Sonic Boom Prediction Workshop, Journal of Aircraft, Vol. 56, No. 3, 2019, pp. 851–875. https://doi.org/10.2514/1.C034866

[7] M. Park, M. Carter, Nearfield Summary and Analysis of the Third AIAA Sonic Boom Prediction Workshop C608 Low Boom Demonstrator, 2021. https://doi.org/10.2514/6.2021-0345

[8] S.S. Stevens, Perceived Level of Noise by Mark VII and Decibels (E), Journal of Acoustic Society of America, Vol 51., 575-601, 1972. https://doi.org/10.1121/1.1912880

Aerospace Science and Engineering - IV Aerospace PhD-Days
Materials Research Proceedings 42 (2024) 132-136

Materials Research Forum LLC
https://doi.org/10.21741/9781644903193-29

Post mission disposal design: Dynamics and applications

Xiaodong Lu[1,a] * and Camilla Colombo[1,b]

[1]Politecnico di Milano, Via La Masa 34, 20156, Milan, Italy

[a]xiaodong.lu@polimi.it, [b]camilla.colombo@polimi.it

Keywords: Luni-Solar Perturbation, Spacecraft End-Of-Life Disposal, Space Debris, Trajectory Design

Abstract. This research addresses post mission disposal design techniques with particular attention to long-term evolution of spacecraft orbits. Semianalytical dynamics models for orbital perturbations are developed to filter out short-periodic perturbation effects and more importantly to improve computational efficiency when dynamics models are integrated in the manoeuvre optimisation problem. The disposal of a spacecraft targeting an Earth re-entry is attained by employing natural orbital perturbations and enhancing natural effects with impulsive manoeuvres. Hamiltonian representations of the system are used to analyse dynamics behaviours and to validate the solutions obtained from optimisation processes.

Introduction

The population of space objects has been increasing more rapidly in last decades, especially with the introduction of mega-constellations. According to European Space Agency (ESA), about 16,990 satellites have been placed into Earth orbit and only about 9000 of them are still functioning. And about 35,150 debris objects are regularly tracked by Space Surveillance Networks (SSN) and maintained in their catalogue [1][2]. The increasing number of space objects in Earth orbits increases probability of collisions between objects leading to a cascade process, known as the Kessler syndrome [3]. In response to this situation, the Inter-Agency Space Debris Coordination Committee (IADC) published space debris mitigation guidelines specifying different mitigation measures [4], one of which is to design end-of-life disposal strategies for spacecrafts [5], preventing prolonged stay in geostationary orbit (GEO) and limiting passage in low Earth orbit (LEO).

A successful end-of-life disposal could make a large contribution to debris mitigation, on the one hand, and implementation of end-of-life could consume large amounts of propellant and significantly increase economic cost, on the other, which decreases feasibility of disposal strategies and discourages spacecraft operators from meeting mitigation guidelines. Therefore, the research aims to develop end-of-life disposal design techniques for spacecrafts, leveraging orbital perturbations to reduce energy consumption during end-of-life disposal.

The research focuses on applying semianalytical propagators to manoeuvre design and optimisation processes, reducing computational burden compared to optimisation with high-fidelity models. The research is composed of three main parts. First, semianalytical models for orbital perturbations are developed based on averaging techniques. This could be implemented in different reference frames with different pros and cons. The following section shows an example in the Earth equatorial frame. Then, the analytical approach to describe orbital perturbations by Kozai using Hamiltonian representations is developed considering J_2 and lunisolar perturbations. This can be exploited in analysing long-term dynamics behaviours and identifying possible disposal strategies qualitatively. It can also be used after the optimised solutions are obtained to validate the effectiveness of the solutions. The last part is to design disposal manoeuvres. This can be done with or without numerical orbit propagation. An optimisation using triple averaged model without numerical orbit propagation is firstly done as a preliminary investigation due to its higher efficiency but less accuracy. In this way, the optimisation require much less computational resources. The obtained results can then be used as a first guess to optimise with numerical orbit propagation using double averaged model to refine the results, since it's more accurate but computationally expensive.

In the next sections, the research performed or planned for each block is explained with the focus on simplification of dynamics models and analysis of dynamics behaviours.

Semianalytical model for orbital perturbations

One of the main difficulties of spacecraft end-of-life disposal design is high computational cost of optimisation process since numerical orbit propagation of decades is involved. Therefore, the main idea is to use semianalytical dynamics models for orbital perturbations to reduce the computational costs and hence the resources needed within the optimisation process.

The dynamics of a spacecraft orbiting the Earth is described as a perturbed two-body problem and is given by the well-known Lagrange planetary equations of motion, where the disturbing function is trigonometric functions of Keplerian elements, periodic in different time scales corresponding to different angles. The idea of semianalytical dynamics model is to separate the short-periodic, long-periodic, and secular perturbation effects to focus on long-term and secular evolution of orbital motion. One method of developing such models is to average the disturbing function over different time intervals to eliminate the fast angles.

The disturbing function of the J_2 perturbation is averaged over one orbital period of a spacecraft [6],

$$\overline{R}_{J_2} = \frac{\mu J_2 R_{\oplus}^2}{4a^3 \eta^3} (2 - 3\sin^2 i), \tag{1}$$

and the one of third-body perturbation is averaged in the same manner,

$$\overline{R}_{3b} = \frac{\mu_3}{r_3} \sum_{l=2}^{\infty} \left(\frac{a}{r_3}\right)^l F_l(A, B, e), \tag{2}$$

and averaged again over one orbital period of the perturbing body,

$$\overline{\overline{R}}_{3b} = \frac{\mu_3}{a_3} \sum_{l=2}^{\infty} \left(\frac{a}{a_3}\right)^l F_l(\alpha_A, \beta_A, \alpha_B, \beta_B, e), \tag{3}$$

where $A, B, \alpha_A, \beta_A, \alpha_B, \beta_B$ are functions of i, Ω, ω of a spacecraft and $e_3, i_3, \Omega_3, \omega_3$ of a perturbing body.

The total single- and double-averaged disturbing functions are following,

$$\overline{R} = \overline{R}_{J_2} + \overline{R}_{Sun} + \overline{R}_{Moon}, \tag{4}$$

$$\overline{\overline{R}} = \overline{R}_{J_2} + \overline{\overline{R}}_{Sun} + \overline{\overline{R}}_{Moon}. \tag{5}$$

To further simplify the dynamics model, one can average the third-body disturbing function over period of variation of RAAN, also known as elimination of the node,

$$\overline{\overline{\overline{R}}}_{3b} = \frac{\mu_3}{a_3} \sum_{l=2}^{\infty} \left(\frac{a}{a_3}\right)^l F_l(e, i, \omega, e_3, i_3, \omega_3), \tag{6}$$

where the node of the third body's orbit is also eliminated since it is coupled with RAAN of a spacecraft.

Aerospace Science and Engineering - III Aerospace PhD-Days
Materials Research Proceedings 42 (2024) 132-136

Materials Research Forum LLC
https://doi.org/10.21741/9781644903193-29

The averaging technique allows one to eliminate fast angles in the disturbing function, hence separating long-periodic and secular effects from the short-periodic ones. This procedure is of importance since it simplifies the manoeuvre optimisation process a lot. The simplified model is validated by comparing with the high-fidelity models and actual ephemerides.

Phase space structures of the dynamics
The Hamiltonian formulation of a system allows to understand the behaviours and qualitative insights of dynamics. Thanks to simplification through averaging, a Hamiltonian formulation of dynamics of orbital perturbations only depending on e, i, ω of a spacecraft is obtained. Furthermore, the so-called Kozai parameter $\Theta = (1 - e^2) \cos^2 i$ [7] is a constant since the z-component of angular momentum is conserved. Hence, one can get a Hamiltonian of one degree-of-freedom and phase space maps of the dynamics [8].

As an example, the left figure of Fig 1 shows a phase space map of e, i, ω of a spacecraft in highly elliptical orbits (HEO) considering lunar perturbation only. It demonstrates that the phase space of the dynamics follows a layer structure where different layer corresponds to different respective values of Kozai parameter. The red line corresponds to $\Theta = 0.3$ while the black line corresponds to $\Theta = 0.1$. The figure of right-hand side is an e, ω map of orbits with a semimajor same as the INTEGRAL mission, and the red curve shows the initial phase curve of the INTEGRAL mission [9]. It is identified that spacecraft orbits evolve under natural perturbations along the phase curve. Therefore, it is possible to enhance natural evolution of orbits by impulsive manoeuvres at some point of a phase curve, driving the orbit to another phase curve and the orbit then evolves under natural perturbations and meets re-entry conditions by entering the critical region after curtain time.

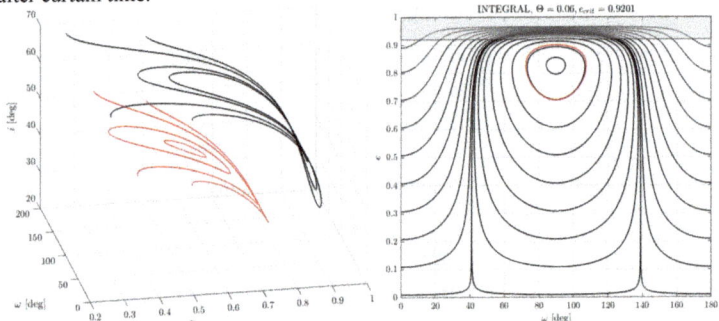

Fig 1 Left: Phase space map, lunar perturbation only, red line:$\Theta = 0.3$, black line: $\Theta = 0.1$; Right: Phase space portrait of the INTEGRAL mission, $\Theta = 0.06, e_{crit} = 0.9201$, red curve: initial phase curve of the INTEGRAL mission.

Disposal manoeuvres design
Following inspiration of the previous section [8], the procedures for computing disposal manoeuvres are following. For any given start point, variations of elements are obtained using Gauss' variational equations, in which Δv is formulated in tangential-normal-subnormal (TNH) frame. The post-manoeuvre elements are propagated using averaged equations to get the maximal eccentricity e_{max}. Assume that a successful re-entry is attained if perigee height is below 100 km, base on which the concept of critical eccentricity e_{crit} is defined. Therefore, the attainment of re-entry is determined by $e_{max} > e_{crit}$.

As a preliminary investigation, a grid search was carried out where magnitudes and directions of manoeuvres are meshed with limits $\Delta v = 100\sim150$ m/s, $\alpha = 0\sim180°, \beta = 0\sim180$. Fig 2 shows the result of the grid search in phase space map and demonstration of manoeuvres

Aerospace Science and Engineering - III Aerospace PhD-Days   Materials Research Forum LLC
Materials Research Proceedings 42 (2024) 132-136   https://doi.org/10.21741/9781644903193-29

corresponding to successful re-entries. The result shows that manoeuvres leading to successful re-entries lie in a cone region along the tangential direction and the most effective ones are the ones in the same or the opposite direction of the tangential direction.

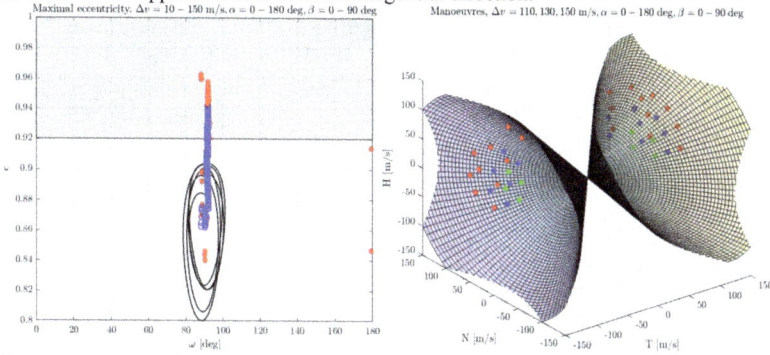

Fig 2 Left: Result of grid search shown in e, ω phase space, black curve: initial phase curve, blue markers: e_{max} for successful re-entries, red markers: e_{max} for non-successful re-entries, grey zone: $e_{max} > e_{crit}$; Right: Demonstration of manoeuvres corresponding to successful re-entries.

Conclusions and next steps

The output of semianalytical models and phase space analysis of dynamics provides possibilities for efficient disposal manoeuvre optimisation. Following the preliminary investigation, the optimisation of disposal manoeuvres with or without numerical orbit propagation can be carried out to identify the optimal solutions. The obtained solutions can then be validated through high-fidelity numerical orbit propagation and through phase space representations. This would be part of the final year research and would focus on the feasibility of analytical approaches.

Acknowledgements

This PhD project is funded by China Scholarship Council (CSC) and has received funding from the European Research Council (ERC) under the European Union's Horizon 2020 research and innovation program (grant agreement No. 679086 – COMPASS).

References

[1] ESA, Space Debris by the Numbers.
https://www.esa.int/Space_Safety/Space_Debris/Space_debris_by_the_numbers

[2] ESA Space Debris Office, ESA's Annual Space Environment Report, 2023.
https://www.sdo.esoc.esa.int/environment_report/Space_Environment_Report_latest.pdf

[3] Kessler, D. J., Cour-Palais, B. G., Collision Frequency of Artificial Satellites: the Creation of a Debris Belt, Journal of Geophysical Research, Vol. 83, No. A6, 1978, pp. 2637-2646.
https://doi.org/10.1029/JA083iA06p02637

[4] IADC, IADC Space Debris Mitigation Guidelines, IADC-02-01, Revision 3, 2021.
https://iadc-home.org/documents_public/file_down/id/5249

[5] Fernández, L.A., Wiedemann, C. & Braun, V. Analysis of Space Launch Vehicle Failures and Post-Mission Disposal Statistics. Aerotecnica Missili & Spazio. 101, 243–256 (2022).
https://doi.org/10.1007/s42496-022-00118-5

[6] Colombo, C., Long-Term Evolution of Highly-Elliptical Orbits: Luni-Solar Perturbation Effects for Stability and Re-entry, Frontiers in Astronomy and Space Sciences, Vol. 6, 2019. https://doi.org/10.3389/fspas.2019.00034

[7] Kozai Y., Secular Perturbations of Asteroids with High Inclination and Eccentricity, The Astronomical Journal, Vol. 67, No. 9, Nov. 1962, pp. 591-598. https://doi.org/10.1086/108790

[8] Lu, X., Colombo, C., Reachable domain analysis for analytical design of end-of-life disposal, in: New Frontiers of Celestial Mechanics: theory and applications, Padova, Italy, 2023.

[9] Colombo C, Letizia F, Alessi EM, Landgraf M. End-of-life Earth re-entry for highly elliptical orbits: the INTEGRAL mission. In the 24th AAS/AIAA space flight mechanics meeting, 2014, pp. 26-30.

Aerospace Science and Engineering - IV Aerospace PhD-Days   Materials Research Forum LLC
Materials Research Proceedings 42 (2024) 137-141   https://doi.org/10.21741/9781644903193-30

Evolutionary predictive model of the space debris environment

Wiebke Retagne[1,a] *, Lorenzo Giudici[2], Camilla Colombo[3]

[1]Ph.D. student, Politecnico di Milano, Via La Masa 52, Milano, Italy

[2]Postdoctoral researcher, Politecnico di Milano, Via La Masa 52, Milano, Italy

[3]Ph.D.Supervisor, Associate Professor, Politecnico di Milano, Via La Masa 52, Milano, Italy

[a]wiebke.retagne@polimi.it

Keywords: Space Safety, Space Debris Modelling, Uncertainty Quantification, Density Approach

Abstract. In recent years, the exponential growth of space debris has become evident. To mitigate debris problem, a precise model for predicting the space debris environment is necessary. This research project tackles this challenge of space debris modelling, through adopting the continuum approach. In the continuum approach a space debris cloud is treated as a fluid. As a novel aspect, the model will include a detailed uncertainty analysis. The challenge here is to find a unified approach to deal with the different uncertainty sources. The analysis will help to identify the largest uncertainty sources and will aid in developing a more precise model. To find a balance between robustness and computational time high performance computing will be employed. Furthermore, the effect of mitigation measures and newly launched missions will be investigated through the combination of historical data with economic forecasting methods, making it possible to make informed decisions for sustainable space operations.

Introduction

In the last 10 years the landscape of space flight has changed drastically. Historically, only national and supranational agencies like NASA [1], ESA [2], Roscosmos [3] were able to put a satellite into space. With companies like SpaceX [4] offering rocket starts commercially, space has become increasingly accessible. In 2021, the start-up sector in space saw a significant increase in funding. Relative to 2020, capital investment grew by 82% [5]. This offers several opportunities, but it also means that the number of players and the number of objects in the popular regions of space like the low-earth orbit and the geostationary orbit is increasing rapidly. Mega-constellations for commercial use by companies such as SpaceX, Amazon [6] and OneWeb [7] further contribute to the rising number of satellites. This sharp increase can be seen in Fig. 1. The number of objects is growing exponentially.

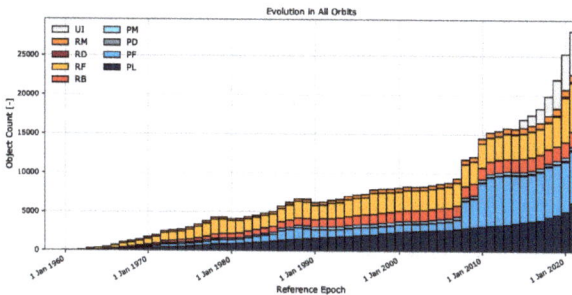

Fig. 1: Evolution of the number of objects in all orbits [8]

Aerospace Science and Engineering - III Aerospace PhD-Days   Materials Research Forum LLC
Materials Research Proceedings 42 (2024) 137-141   https://doi.org/10.21741/9781644903193-30

As a consequence, space debris has become an increasingly relevant topic. Space debris refers to any artificial object in space that is no longer functioning. This also includes fragments and any parts thereof. Space debris poses a risk to active satellites, since even small fragments could cause fatal damage [9]. In 2009 a catastrophic collision occurred between the non-functioning satellite Cosmos 2251 and the operational satellite Iridium 33. This collision created a large number of fragments, that span multiple altitudes. The collision was caused by lack of knowledge about the position of the Cosmos 2251 satellite [10]. In the worst-case scenario, collisions like these could cause a chain effect, where the fragments created by one collision collide with other objects and so on. This is called the Kessler syndrome [11]. To keep space usable for future generations, space debris has to be mitigated and collisions have to be avoided. For this purpose, an accurate model of the space debris population has to be developed.

This research is part of the GREEN SPECIES project, funded by the European Research Council on the "Robust control of the space debris population to define optimal policies and an economic revenue model for sustainable development of space activities" and aims at developing a highly accurate and robust space debris model, including an uncertainty analysis.

State of the Art

The modelling of space debris is usually separated into two main approaches: A deterministic and a statistical approach. In the deterministic approach each space debris object is tracked and propagated individually. This approach is widely used and assesses the biggest collision risks [12] Fragments smaller than 10 cm are usually neglected in these simulations, since the computational cost associated with modelling each small fragment is exceedingly high. The growing number of objects in space and the increasingly high traffic neglecting centimeter or sub-centimeter debris results in underestimating the actual threat for safe in-orbit operations. Therefore, a comprehensive model of these smaller fragments is necessary for safe and sustainable use of space.

Instead of deterministic models, a continuum approach can be adopted for these smaller fragments. The continuum approach treats the small space debris objects as a debris cloud, which is modelled through a density function. This density function can be propagated using the equation of continuity [13]

$$\frac{\partial n}{\partial t} + \nabla \cdot (nf) = \dot{n}^+ - \dot{n}^-$$

where n is the phase space density, f represents the forces that account for the orbital dynamics and orbit perturbations and $\dot{n}^+$ and $\dot{n}^-$ account for the sources and sinks of space debris objects, respectively. Solving this equation is not trivial, since it represents a Partial Differential Equation (PDE).

The first fully analytical solution for the long-term propagation of debris in Low Earth Orbit (LEO) was presented by McInnes in 1993 [13] based on the Method of Characteristics (MoC). The MoC transforms the PDE into a system of ordinary differential equations and the solution is obtained along characteristic lines. McInnes solved the continuity equation in two dimensions: Time and the radial distance from the center of the Earth. As a perturbation only the atmospheric drag was considered. In recent years, Letizia et al. (2015) [14] has built on the work of McInnes and adopted the approach to model the long-term evolution of a debris cloud generated as a result of a fragmentation event. This method was extended into multiple dimensions (Letizia et al. (2016) [15]). With the work of Colombo et al. (2016) [16] a multi-dimensional approach for the propagation of the global space debris population was presented.

Instead of solving the equation analytically the characteristics can also be numerally propagated. Frey, Colombo et al. (2019) [17] presented the Starling suite, which estimates the non-linear evolution of densities in orbit. A Gaussian Mixture Model (GMM) is applied to the numerically propagated characteristics in order to obtain a simpler, surrogate model. The solution is also extended into the multi-dimensional domain.

In Giudici et al. (2023) [18], the Starling suite was extended to model any non-linear dynamics. Another approach to solve this problem is the finite volume method. This describes the process of discretizing the space into finite volumes and propagating the density using the continuity equation for each of these volumes. The equations are integrated numerically. This has the benefit of immediately achieving the full density [19]. These two models were combined and used to develop COMETA for the long-term propagation of the debris environment in LEO using the continuum approach [20].

To prevent the worst-case scenario of a cascading effect from happening, the space debris environment needs to be well understood. This not only includes modelling the existing space debris environment, but also requires an understanding of how the environment develops in the future. The main goal of this research project is to provide a better understanding of the space debris environment and the uncertainties associated with it. This should also include the effect different events have on it as a whole.

Methodology

In this research, the model of the space debris environment will be extended to include uncertainties to include uncertainties. For this the continuum approach described previously will be adopted. The main research goal of this PhD is to identify and address the biggest sources of influence in the model.

Dolado et. al (2015) [21] presented a classification of uncertainty sources in space debris modelling. This will be built upon to classify uncertainties based on the biggest impact on the space debris environment.

Uncertainties not only stem from numerical modelling, but also from socio-economic and political circumstances. One example of this kind of uncertainty stems from the launch traffic model.

While the actual launch traffic is out of control for the modeler, the model itself can be improved upon. Most models, such as DELTA from ESA [22] or LEGEND from NASA [23] simply repeat the historical launch pattern, while treating constellations separately. This research instead aims to implement a highly adaptable and accurate launch traffic model by combining the study of historical data with economic and financial forecasting models. Additionally, post-mission disposal failures influence the debris environment. Studying these as done in Fernándenz et. al [24] and quantifying them will help improve the space debris model.

Next to these socio-economic uncertainties, the space debris environment is also highly influenced by physical phenomena such as e.g. solar activity and the atmospheric drag. The strength of the atmospheric drag influences the lifetime of a space debris object in orbit. The solar cycle influences the atmospheric drag, showing that uncertainties in the space debris model are also highly dependent on each other. While for the launch traffic model it was possible to treat it separately from the other uncertainties, here both play a role together. An approach needs to be found that can treat the interacting uncertainties concurrently. For the uncertainty in parameters like the area-to-mass ratio an uncertainty distribution could be defined, which can be treated as a density. The density equation would evolve with a zero derivative. This would allow the uncertainty to be propagated with the continuity equation directly. For the uncertainties of the dynamic equations a more complex approach needs to be defined.

For this purpose, several different uncertainty quantification methods will be investigated. In many applications Monte Carlo methods are used to quantify uncertainties. This could quickly become too computationally expensive for the space debris model. Therefore, other methods such as polynomial chaos expansion or propagating a probability density function directly will be investigated. In Trisolini et. al (2021) [25] the uncertainties of the re-entry dynamics were

Aerospace Science and Engineering - III Aerospace PhD-Days   Materials Research Forum LLC
Materials Research Proceedings 42 (2024) 137-141   https://doi.org/10.21741/9781644903193-30

propagated directly in combination with the dynamics with the continuity equation. It will be investigated whether this approach can also be used for the global space debris environment. This would include extending the current 8-dimensional model to higher dimensions to add the uncertainty as a dimension. Here, this could be either done through the MoC or the FVM approach. Next to the propagation of uncertainties, the model can also be validated using existing data. Horstmann et. al (2019) [26] formulated an error function to classify uncertainties in the deterministic model. To improve computational time surrogate models could also be employed. In the end a balance needs to be found between the computational complexity and the accuracy of the model. Using high performance computing will also be explored. Based on this the best method will be selected.

Conclusions

This research project will give the framework for sustainable use of the space environment. With the inclusion of uncertainties into the space debris modelling, the models can move on from sensitivity analysis to accurate forecasting models. Including the effect of mitigation strategies and newly launched missions will help in guiding future policy decisions and will ensure safe and sustainable space operations in the future.

Acknowledgements

The research presented in this thesis received funding from the European Research Council (ERC) under the European Union's Horizon Europe research and innovation program as part of the GREEN SPECIES project (Grant agreement No 101089265).

References

[1] "NASA." Aug. 2023. Accessed: Mar. 13, 2024. [Online]. Available: https://www.nasa.gov/

[2] "ESA." Aug. 2023. Accessed: Mar. 13, 2024. [Online]. Available: https://www.esa.int

[3] "Roscosmos." Mar. 2024. Accessed: Mar. 13, 2024. [Online]. Available:
http://archive.government.ru/eng/power/106/

[4] "SpaceX." Aug. 2023. Accessed: Mar. 13, 2024. [Online]. Available:
https://www.spacex.com/

[5] Bryce Tech, "Start-Up Space: Update on investment in commercial space ventures," Aug. 2022. Accessed: Mar. 13, 2024. [Online]. Available: https://brycetech.com/reports

[6] "Amazon Mega Constellation Project." Aug. 2023. Accessed: Mar. 13, 2024. [Online]. Available: https://www.aboutamazon.com/news/innovation-at-amazon/what-is-amazon-project-kuiper

[7] "OneWeb." Aug. 2023. Accessed: Mar. 13, 2024. [Online]. Available: https://oneweb.net/

[8] E.S.D Office, "ESA's annual space environment report," 2023.

[9] G. Drolshagen, "Impact effects from small size meteoroids and space debris," *Advances in Space Research*, vol. 41, no. 7, pp. 1123–1131, 2008. https://doi.org/10.1016/j.asr.2007.09.007

[10] C. Pardini and L. Anselmo, "The short-term effects of the cosmos 1408 fragmentation on neighboring inhabited space stations and large constellations," *Acta Astronaut*, vol. 210, pp. 465–473, 2023. https://doi.org/10.1016/j.actaastro.2023.02.043

[11] D. J. Kessler and B. G. Cour-Palais, "Collision frequency of artificial satellites: The creation of the debris belt," *J Geophys Res Space Phys*, vol. 83, no. A6, pp. 2637–2646, 1978. https://doi.org/10.1029/JA083iA06p02637

Aerospace Science and Engineering - III Aerospace PhD-Days
Materials Research Proceedings 42 (2024) 137-141

Materials Research Forum LLC
https://doi.org/10.21741/9781644903193-30

[12] S. Flegel *et al.*, "The master-2009 space debris environement model," *European Space Agency, (Special Publication) ESA SP*, vol. 672, Mar. 2009.

[13] C. R. McInnes, "An analytical model for the catastrophic production of orbital debris," *ESA J*, vol. 17, no. 4, pp. 293–305, 1993.

[14] F. Letizia, C. Colombo, and H. G. Lewis, "Analytical model for the propagation of small-debris object cloud after fragmentations," *Journal of Guidance, Control, and Dynamics*, vol. 38, no. 8, pp. 1478–1491, 2015. https://doi.org/10.2514/1.G000695

[15] F. Letizia, C. Colombo, and H. G. Lewis, "Multidimensional extension of the continuity equation method for debris cloud evolution," *Advances in Space Research*, vol. 57, no. 8, pp. 1624–1640, 2016. https://doi.org/10.1016/j.asr.2015.11.035

[16] C. Colombo, F. Letizia, and H. G. Lewis, "Spatial density approach for modelling of the space debris population," in *26th AAS/AIAA Sp. Flight Mech. Meet*, 2016, pp. 16–465.

[17] S. Frey, C. Colombo, and S. Lemmens, "Application of density-based propagation of fragment clouds using the Starling suite," in *1st International Orbital Debris Conference (IOC)*, Sugar Land (TX, USA), 2019, pp. 1–10. https://doi.org/10.1016/j.asr.2019.03.016

[18] L. Giudici, M. Trisolini, and C. Colombo, "Probabilistic multi-dimensional debris cloud propagation subject to non-linear dynamics," *Advances in Space Research*, vol. 72, no. 2, pp. 129–151, 2023. https://doi.org/10.1016/j.asr.2023.04.030

[19] L. Giudici and C. Colombo, "Space debris density propagation through a finite volume method," Mar. 2023.

[20] L. Giudici, C. Colombo, A. Horstmann, F. Letizia, and S. Lemmens, "Density-based evolutionary model of the space debris environment in low-Earth orbit," *Acta Astronaut*, vol. 219, pp. 115–127, Jun. 2024. https://doi.org/10.1016/j.actaastro.2024.03.008

[21] J. Dolado, C. Pardini, and L. Anselmo, "Review of uncertainty sources affecting long-term predictions of space debris evolutionary models," *Acta Astronaut*, vol. 113, 2015. https://doi.org/10.1016/j.actaastro.2015.03.033

[22] B. B. Virgili, "DELTA (DEBRIS ENVIRONMENT LONG-TERM ANALYSIS)," 2016. [Online]. Available: https://api.semanticscholar.org/CorpusID:197866235

[23] J.-C. Liou, D. Hall, P. Krisko, and J. N. Opiela, "LEGEND – a three-dimensional LEO-to-GEO debris evolutionary model," *Advances in Space Research*, vol. 34, pp. 981–986, Mar. 2004. https://doi.org/10.1016/j.asr.2003.02.027

[24] Fernández, L. A., Wiedemann, C., & Braun, V. (2022). Analysis of Space Launch Vehicle Failures and Post-Mission Disposal Statistics. Aerotecnica Missili & Spazio, 101(3), 243–256. https://doi.org/10.1007/s42496-022-00118-5

[25] M. Trisolini and C. Colombo, "Propagation and Reconstruction of Reentry Uncertainties Using Continuity Equation and Simplicial Interpolation," *Journal of Guidance, Control, and Dynamics*, vol. 44, no. 4, pp. 793–811, Apr. 2021. https://doi.org/10.2514/1.G005228

[26] A. Horstmann, H. Krag, and E. Stoll, "Providing Flux Uncertainties in ESA-MASTER: The Accuracy of the 1cm Population," in *First International Orbital Debris Conference*, in LPI Contributions, vol. 2109. Dec. 2019, p. 6015.

Aerospace Science and Engineering - IV Aerospace PhD-Days
Materials Research Proceedings 42 (2024) 142-145

Materials Research Forum LLC
https://doi.org/10.21741/9781644903193-31

Model based systems engineering and concurrent engineering for space systems

Alessandro Mastropietro[1,2,a] *

Kongsberg NanoAvionics UAB, Mokslininku str. 2A, LT-08412 Vilnius, Lithuania

[2]University of the Bundeswehr Munich, Werner-Heisenberg-Weg 39, 85579 Neubiberg, Germany

[a]alessandro.mastropietro@nanoavionics.com

Keywords: New Space, Model Based System Engineering, Concurrent Engineering, Fractionated and Federated Systems

Abstract. The objective of this paper is to present the rationale and application of Model Based System Engineering and Concurrent Engineering for space systems. In the context of the New Space Economy, characterized by disruptive technologies, standardization, and the entry of private investments into the industry, it is critical to manage space systems effectively among the different involved stakeholders. In order to address all related challenges, systems modelling approaches are used at different levels, from subsystems to satellites to complex mission architectures, like fractionated and federated systems.

Introduction

The New Space Economy refers to the growing commercialization of space services and exploration driven by disruptive technologies and increased accessibility. Private investment is surging, and Morgan Stanley predicts the global space industry could reach $1 trillion by 2040 (Fig. 1), impacting various sectors. Satellite broadband Internet access is expected to contribute significantly to this growth [1].

In this dynamic landscape, the juxtaposition of standardization and disruptive technologies presents a central challenge. Standardization, as demonstrated by the success of CubeSat technology, offers cost-effective development, while disruptive technologies, exemplified by Inter Satellite links, continuously reshape the market. Harmonizing these aspects becomes crucial to swiftly deliver new services, demanding adaptability across various systems while minimizing redesign costs. Effective trade-offs and design change impact analyses play a pivotal role in navigating this intricate balance and arriving at optimal solutions.

The complexity of space programs has led to the study of Model-Based Systems Engineering (MBSE) and Concurrent Engineering (CE) approaches for efficient data management and collaboration across disciplines. MBSE has demonstrated effectiveness in small-sat lifecycle management [2], while CE is a well-proven approach for designing complex space systems in the pre-phase 0/A [3].

The current elaborate will present these methodologies, the scenario of a complex space mission and the implementation steps to arrive at a modelling tool for distributed systems design management.

Methodologies

MBSE approaches have been applied in different engineering domains. They enhance the ability to capture, analyse, share and manage the information associated with a system [4]. Among the various solutions in terms of tool and methodology, the Capella tool (Fig. 1), based on the Arcadia [5] methodology has been selected in the frame of the doctoral project. Capella supports system engineering activities at different levels: operational analysis, system analysis, logical architecture, physical architecture and EPBS (End Product Breakdown Structure).

Concurrent engineering is a method of designing and developing products, which overcomes

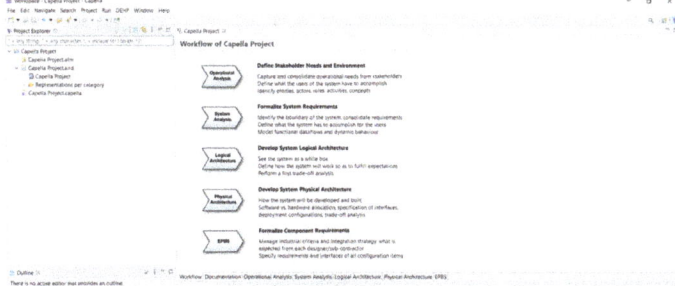

Fig. 1 Capella Workspace

the more classic sequential and centralized design approaches. Different expertise teams work simultaneously communicating with each other. Benefits can be seen in performance, with reduced study duration, reduced costs, and quality improvement. A typical concurrent engineering study process is characterized by a team of specialists working in a dedicated facility for different sessions using a centralized system data model to share information. Regarding the latter, the COncurrent Model-based dEsign Tool (COMET) (Fig. 2) [6] is the chosen software which enables the different members of a team to co-work on the same system and exchange design information through different iterations.

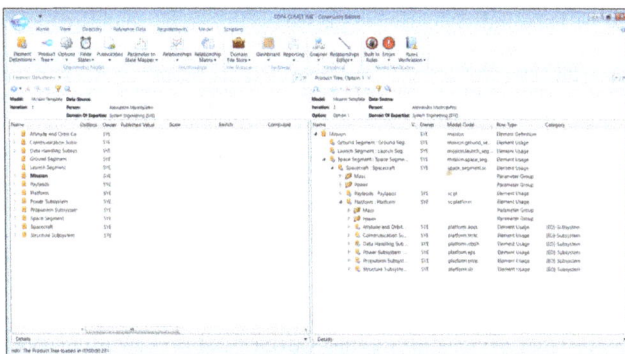

Fig. 2 COMET Workspace – Mission Template

MBSE and CE support the centralization of design information, acting as a source of truth. This empowers decision-makers with a comprehensive understanding, potentially accelerating Time to Market (TTM) [7]. The methodologies establish traceability for requirements and functionalities across system architecture levels, ensuring alignment with stakeholder expectations and facilitating adaptive responses to evolving needs. The capacity to analyse the impact of design changes on system architecture, along with RAMS analysis, enhances resilience and fosters operational confidence. Furthermore, this modelling approach contributes to the realization of a digital twin and, consequently, to the improvement of a model, development after development. This dynamic approach proves vital in accurately assessing and analysing the ever-evolving landscape of New Space, where the entry of new players intensifies the importance of safety, reliability, and adaptability considerations in both hardware and software domains.

Fractionated and Federated Systems

Over the past decade, increasing global demand for connectivity, navigation services and climate monitoring has caused a profound transformation of the space market [8]. This change has been facilitated by recent technological advances, which have introduced higher levels of digitization, miniaturization and reusability. [9].

The physical and service limitations of the classic space infrastructures, characterized by single big platforms, highlight the potential of a multi-mission scenario ecosystem of fractionated and federated space systems for future space missions [10]. Fractionation, involving the distribution of functions among interconnected modules, and federation, clustering satellites to dynamically share resources, are collaborative approaches to enhance overall satellite network performance.

The design complexities encompass various aspects, from architecture definition and payload design to co-design of space and ground segments, demanding efficient decision-making and facilitating easier maintainability for future development. Emphasizing the critical role of effective stakeholder communication, safety, and reliability considerations, it's necessary to meticulously track design decisions for continuous improvement in future space mission development, ultimately optimizing resource utilization.

Implementation

Contributing to the realization of a proof-of-concept digital tool able to support federated and fractionated space systems design, like a multi-layer constellation for telecommunication purposes, requires a solid and robust foundation. The starting point for this work is the assessment of already existing tools with a forward-looking vision. Modelling spacecraft subsystems allows training on the modelling tools and methodology showing its strengths, to be exploited, and weaknesses, to be improved. Subsequently, a spacecraft can be considered to experience the concept of a System of Systems (SoS) with respect to its subsystems. The design of a payload and its integration into the spacecraft presents here the perfect opportunity to develop co-design models and study their maintainability. At this level, it is already possible to make important considerations about the best strategies for co-engineering from a modelling point of view and try to understand which level of abstraction is needed depending on the mission ecosystem and stakeholders. Finally, the modelling of a complex space mission scenario can be faced. Many challenges shall be addressed from the design of a distributed space segment to its co-design with the ground segment. The experience gained at the different levels of research will make it possible to make the best use of modelling capabilities and, if necessary, to propose new modelling solutions.

Conclusion

Integrating Model-Based Systems Engineering (MBSE) and Concurrent Engineering (CE) has proven to be promising in space applications, offering potential benefits for managing space systems architectures, improving decision-making, and fostering collaboration. However, implementing these methodologies coherently within the New Space sector poses challenges in aligning diverse stakeholders, standardization and disruptive technologies. In the context of the MSCA doctoral program Harmony [10], the objective of the current research is to develop a digital tool, based on MBSE and CE, to aid decision-makers in designing space systems, emphasizing understanding parameter impacts and interface changes for robustness and adaptability. Moreover, the consortium's multidisciplinary nature facilitates training on model-based and concurrent engineering approaches for distributed space systems and their implementation.

Acknowledgements

This study was conducted as part of the MSCA Industrial Doctorate program HARMONY: Innovating New Space Frontiers: Harmonised Federated And Fractionated Systems Unlocking Fresh Perspectives For Satellite Services, funded by the European Union's Horizon Europe research and innovation program under grant agreement N°101072798.

References

[1] Information on https://www.morganstanley.com/ideas/investing-in-space

[2] P. Minacapilli, M. Lavagna, Small sats lifecycle management through MBSE aided decision making tailored tool, 72nd International Astronautical Congress (IAC 2021), (2021).

[3] J. Hoffmann, M. Deutsch, R. Bertrand, Development of a Concurrent Engineering Tutorial as part of the "ESA_Lab@" initiative, 4th Symposium on Space Educational Activities, Barcelona, (2022). https://doi.org/10.5821/conference-9788419184405.029

[4] L. Bitetti, R. De Ferluc, D. Mailland, G. Gregoris, F.Capogna: Model-Based Approach for RAMS Analyses in the Space Domain with Capella Open-Source Tool, International Symposium on Model-Based Safety and Assessment, (2019). https://doi.org/10.1007/978-3-030-32872-6_2

[5] P. Roques, Systems Architecture Modeling with the Arcadia Method, first ed., ISTE Press Ltd and Elsevier Ltd, UK, 2018. ISBN 9781785481680. https://doi.org/10.1016/B978-1-78548-168-0.50001-3

[6] RHEA GROUP, CDP4-COMET A platform for Collaborative Model Based System Engineering User Manual, (2024)

[7] Information on https://www.tcgen.com/time-to-market/

[8] Information on https://www.unoosa.org/oosa/osoindex/index.jspx?lf_id=

[9] O. Kodheli, E. Lagunas, N. Maturo, S.K. Sharma, B. Shankar, D. Spano, S. Chatzinotas, S. Kisseleff, J. Querol, L. Lei, T.X. Vu, G. Goussetis, Satellite Communications in the New Space Era: A Survey and Future Challenges, IEEE Comms Surveys & Tutorials, Vol. 23, No. 1, pp. 70-109, (2021). https://doi.org/10.1109/COMST.2020.3028247

[10] Information on https://www.harmony-horizoneurope.eu/

Aerospace Science and Engineering - IV Aerospace PhD-Days
Materials Research Proceedings 42 (2024) 146-149

Materials Research Forum LLC
https://doi.org/10.21741/9781644903193-32

Cislunar orbit dynamics with an application to space situational awareness

Mathilda Bolis[1,a] *, Elisa Maria Alessi[2,b] and Camilla Colombo[1,c]

[1]Department of Aerospace Science and Technology, Politecnico di Milano, Italy

[2]Istituto di Matematica Applicata e Tecnologie Informatiche "Enrico Magenes", CNR, Italy

[a]mathilda.bolis@polimi.it, [b]em.alessi@mi.imati.cnr.it, [c]camilla.colombo@polimi.it

Keywords: Cislunar Space, Periodic Orbits, Stability, End-Of-Life, Disposal, Dynamical Systems

Abstract. Given the growing number of planned missions and satellite launches, it is clear the cislunar orbital domain gained great interest from both industries and space agencies in the last few years. The chaotic dynamics that characterise this area of space set new challenges that have not been considered for near-Earth space. An increasing number of space objects orbiting in the chaotic cislunar domain could lead to a more complex debris problem than is found near-Earth today, making the design of effective End-of-Life (EoL) disposal strategies more important than ever. This Ph.D. research aims to develop a dynamic cartography of cislunar space through tools derived from dynamical systems theory. Thus, it will be possible to characterise the long-term stability of different families of orbits in cislunar space, creating a catalogue of stable and unstable orbits; to develop a tool to define effective and customisable EoL strategies; and to analyse the behaviour of possible fragmentations in the cislunar domain.

Introduction

Space is one of the final frontiers of scientific exploration: its importance as an economic and strategic asset has never been more pronounced than now. In a context in which hundreds of space missions are planned for the next few years and the number of in-orbit satellites exponentially increases, both industries and space agencies have shown a strong interest in the cislunar domain, a region strongly influenced by both the Earth and the Moon. The dynamics of this region cannot be modelled with the techniques developed for near-Earth space. The influence of the Moon, other celestial bodies, and multiple perturbations make the motion of an object in this region inherently chaotic. This means that, given the initial state of a satellite, it is not trivial to determine what its final state will be, especially when long-term simulations are considered. Consequently, if the number of space objects in the cislunar region increases, as has been observed lately in the vicinity of Earth [1], a potential debris issue will arise there as well, with additional complications given by the chaotic dynamics. In this context, one of the most effective mitigation measures is the careful design of the EoL phase of a mission. Creating a method to achieve simple and customizable disposal strategies is a matter of primary importance.

This Ph.D. research will therefore focus on the development of a dynamic cartography of cislunar space, thereby trying to define, through dynamical systems theory, parameters that will help characterize the state of a spacecraft. This will make it possible to develop, as a result, a tool capable of selecting the best disposal strategy for a spacecraft, given as input its operational orbit, end-of-life epoch, and amount of fuel still on board. Ideally, the legal aspects of the problem will also need to be considered in the development of this tool. The following sections present a description of the methodologies to be developed as well as of the planned activities.

Aerospace Science and Engineering - III Aerospace PhD-Days   Materials Research Forum LLC
Materials Research Proceedings 42 (2024) 146-149   https://doi.org/10.21741/9781644903193-32

Dynamic models

Many dynamical models can at least partially describe cislunar space dynamics. The Circular Restricted Three-Body Problem (CR3BP) [2] provides a good approximation of the qualitative behaviour of the motion of a spacecraft when its dynamic is mainly governed by the influence of two primaries with nearly circular orbits and a small inclination difference. Within it, it is possible to identify one energy integral of motion, the Jacobi Constant (JC), and five equilibrium points of the system, the Lagrangian points [3].

When considering the cislunar region, the main perturbation which is not included in the earlier model is the Sun's gravitational pull. To take it into account, the CR3BP can be modified into a Bicircular Restricted Four-Body Problem (BR4BP) (e.g. [4]). The major advantage of using this model is that, in general, it is simple to handle, implement, and use, although the effect of a third major body is added to the system. Moreover, despite in this case the motion of the three main bodies is not coherent, the model shows behaviours qualitatively similar to the ephemeris one, especially in some areas of the phase space.

The dynamic model that most closely matches reality is the N-body problem [5], which, however, comes with a greater computational effort. The influence of perturbations also plays a significant role in determining the motion of a spacecraft in cislunar space. For example, the effect of Solar Radiation Pressure (SRP) [6] becomes particularly relevant in the case of spacecraft with large area-to-mass ratios, while the lunar gravitational field influence has a not negligible effect when in proximity of the Moon [7].

Families of orbits in cislunar space

A particularly interesting aspect of CR3BP is that, by leveraging some of its properties, it is possible to define families of periodic and quasi-periodic orbits. Once initial conditions for possible periodic orbits are identified, differential corrections are exploited to refine them, and continuation methods are used to compute different orbits belonging to the same family [3]. These orbits can then be refined in more complex dynamical models. Families of periodic orbits can be found in cislunar space through continuation techniques, such as are Lyapunov orbits, Halo orbits, Near Rectilinear Halo orbits (NRHOs), Lissajous Orbits, Distant Retrograde orbits, etc. [8][9]. Cycler or horseshoe orbits (e.g. [10]), as well as orbits around the Moon, equivalent, for example, to Highly Elliptical Orbits (HEOs) around the Earth, can also be observed.

End-of-life phase design for spacecraft in cislunar space

The EoL phase of a mission is designed by performing disposal primarily with one of the following four strategies: insertion into a heliocentric orbit, impact on the Moon, destructive Earth re-entry, or insertion into a cislunar frozen graveyard orbit. Among these, the most complex, least studied, and exploited approach is the last one.

In [11], all four approaches are considered, and a dynamic cartography of cislunar space based on Monte Carlo (MC) simulations is proposed. Samples are generated along some cislunar orbits chosen as case studies and, after applying a manoeuvre to each of them, a random perturbation is introduced into the system and several MC simulations are performed, thus evaluating the robustness of each solution to possible disturbances. An analysis is also performed about the possibility for the satellite to interfere with the geostationary pseudo-region.

In [12] instead, the EoL design for GAIA, orbiting Sun-Earth L_2 (SEL2), is formulated as a multi-objective optimization problem. This approach is proposed for disposal with impact on the Moon, with destructive re-entry to Earth, and with insertion into a heliocentric orbit. For the latter case, an energy approach is also suggested, based on the variation of the problem JC to close the Hill curves in SEL2. An alternative approach, based on defining manoeuvres that ensure the constant increase of the minimum orbit intersection distance, the minimum distance between Earth and the disposal orbit, is proposed in [13]. Some of the cited studies refer to the design of the EoL phase of missions orbiting Lagrangian points of the SE system. Nevertheless, it is considered of interest to cite their contributions, as some strategies developed in those contexts could also be applied to cislunar space. Following the same logic, the case reported in [14] is also analysed. Possible disposal options at EoL for HEOs around Earth are studied, taking advantage of natural orbital perturbations and phase space to plan manoeuvres that ensure low fuel consumption.

Methodology and planned activities

Given what was discussed in the previous sections, this research work will be organized as outlined in the following paragraph.

First, a customized propagator for cislunar space will be developed, where the effect of perturbations will be included in the dynamic model only if considered truly non-negligible, exploiting both theoretical astrodynamics and long-term computer simulations. This will give the possibility to combine the simplicity of the CR3BP with the precision of an N-body perturbed model. Data recovered from past missions will be used to validate the developed tool.

It will be interesting then to define a catalogue of cislunar families of orbits, highlighting their characteristics and possible uses for missions with different objectives. An analysis of the stability properties of these families of orbits will follow. To do it, it will be necessary to identify stable and unstable areas of the phase space, exploiting tools such as Poincaré maps, finite-time Lyapunov exponents, and other tools of dynamical systems theory. The analyses developed will also be applied to the study of possible fragmentations in cislunar space.

Properties of HEO with large semi-major axes and eccentricities will be analysed, thanks to Keplerian map theory. In addition, a possible extension of the same approach to HEOs around the Moon and Lagrangian point orbits will be proposed. Also, strategies such as the ones developed in [14] can be exploited for the same application.

Finally, accurate design of EoL for cislunar missions will be performed, highlighting not how the single EoL scenario can be applied to the single mission, but identifying areas of space where a certain disposal strategy proves to be the optimal option, in terms of fuel savings, duration of the disposal phase, and possible risks to current or future missions. Passive disposal solutions, designed by taking advantage of the chaotic dynamics of cislunar space, would be preferred. Similar results have been obtained in the context of the ReDSHIFT project [15], for a region extending from low Earth orbit to the geostationary orbit.

The knowledge developed in this context could also be applied to the design of transfer trajectories and operational orbits for different missions, identifying frozen and quasi-frozen orbits or conditions of special interest.

As an outcome, an algorithm to select the best disposal strategy for a spacecraft will be developed, given as input its operational orbit, EoL epoch and amount of fuel remaining on board, avoiding performing for each disposal design hundreds of numerical simulations. Ideally, the algorithm will need to also consider the legal aspects of the problem, to represent, as accurately as possible, the complexities that characterize the cislunar domain. The developed tool will then be applied to specific cases, possibly ESA missions.

Conclusions

The research introduced in this paper focuses on the development of a dynamical cartography of cislunar space using tools derived from dynamical systems theory. This research aims to create a catalogue of orbit families in cislunar space, to study their stability properties, and to develop a tool useful for defining effective and customizable EoL disposal strategies.

First, dynamical models that are well suited to cislunar space are briefly described and presented, along with the periodic and quasi-periodic orbits that characterize it. Then a brief overview of the methods developed for disposal at EoL is given. Finally, planned activities within this Ph.D. research are described and briefly commented on.

Acknowledgments

This research has received funding as part of the work developed for the agreement n. 2023-37-HH.0 for the project "Attività tecnico-scientifiche di supporto a C-SSA/ISOC e simulazione di architetture di sensori per SST", established between the Italian Space Agency (ASI) and Politecnico di Milano (POLIMI).

References

[1]   ESA's Annual Space Environment Report, 2021. URL: https://www.sdo.esoc.esa.int/environment_report/Space_Environment_Report_latest.pdf

[2]   Szebehely, V. Theory of Orbit: The restricted problem of three Bodies. Academic Press, New York, 1967. doi:10.1016/B978-0-12-395732-0.X5001-6.

[3]   Koon, W. S., Lo, M. W., Marsden, J. E. and Ross, S. D. Dynamical Systems, the Three-Body Problem and Space Mission Design. Marsden Books, 2011.

[4]   Mingotti, G. P. Trajectory Design and Optimization in Highly Nonlinear Astrodynamics. PhD thesis, Politecnico di Milano, Milan, Italy, 2010.

[5]   Meyer, K., Hall, G., and Offin, D. Introduction to Hamiltonian Dynamical Systems and the N-Body Problem, Second Edition, Vol. 90. Springer New York, 2009. https://doi.org/10.1007/978-0-387-09724-4

[6]   Jorba-Cuscó, M., Farrés, A., and Jorba, À. On the stabilizing effect of Solar Radiation Pressure in the Earth-Moon system. Advances in Space Research, 67(9), 2812–2822, 2021. https://doi.org/10.1016/j.asr.2020.01.017

[7]   Michael, W.H., Blackshear, W.T. Recent results on the mass, gravitational field and moments of inertia of the moon. The Moon 3, 388–402 (1972). https://doi.org/10.1007/BF00562460

[8]   Howell, K. C. Three-dimensional, periodic, 'Halo' orbits. Celestial Mechanics 32, 53–71, 1984.

[9]   Hénon, M. Generating Families In The Restricted Three-Body Problem. Springer-Verlag, Springer Science Business Media, 1997.

[10]   De Leo, L., Pontani, M. Low-Thrust Orbit Dynamics and Periodic Trajectories in the Earth–Moon System. Aerotec. Missili Spaz. 101, 171–183 (2022). doi:10.1007/s42496-022-00122-9

[11]   Guardabasso, P., Skoulidou D. K., Bucci L., Letizia F., Lemmens S. and Lizy-Destrez S. Cislunar debris mitigation development of a methodology to assess the sustainability of lunar missions. in 72nd International Astronautical Congress (IAC), Dubai, 2021.

[12]   Armellin, R., Rasotto, M., Di Lizia, P. and Renk, F. End-of-life disposal of libration point orbit missions: The case of Gaia. Advances in Space Research 56, 461–478, 2015. https://doi.org/10.1016/j.asr.2015.03.014

[13]   Alessi, E. M. and Sánchez, J.-P. MOID-increasing disposal strategies for LPO missions. in 65th International Astronautical Congress (IAC), Toronto, 2014.

[14]   Colombo C., Alessi E. M., Van der Weg W., Soldini S., Letizia F., Vetrisano M., Vasile M., Rossi A., Landgraf M. End-of-life disposal concepts for Libration Point Orbit and Highly Elliptical Orbit missions. Acta Astronautica, Vol. 110, p. 298-312, 2015. https://doi.org/10.1016/j.actaastro.2014.11.002

[15]   Rossi A., Colombo C., Tsiganis K., Beck J., Rodriguez J. B., Walker S., Letterio F., Dalla Vedova F., Schaus V., Popova R., Francesconi A., Stokes H., Schleutker T., Alessi E. M., Schettino G., Gkolias I., Skoulidou D. K., Holbrough I., Bernelli Zazzera F., Stoll E. and Kim Y. ReDSHIFT: A global approach to space debris mitigation. Aerospace 5, 2018. https://doi.org/10.3390/aerospace5020064

Aerospace Science and Engineering - IV Aerospace PhD-Days
Materials Research Proceedings 42 (2024) 150-153

Materials Research Forum LLC
https://doi.org/10.21741/9781644903193-33

Vision-based relative navigation system for autonomous proximity orbital operations

Matteo Forasassi[1,a*], Giordana Bucchioni[1,b], Lorenzo Pollini[1,c]

[1]Department of Information Engineering, University of Pisa, Via G. Caruso 16,56122, Pisa, PI Italy

[a]matteo.forasassi@phd.unipi.it, [b]giordana.bucchioni@unipi.it, [c]lorenzo.pollini@unipi.it

Keywords: Autonomous Systems, Pose Estimation, Rendezvous, Vision-Based, CNN, Sounding Balloon, GNC

Abstract. The paper presents the research project of the author, focused on developing, implementing and testing a vision based relative navigation system for spacecraft. A temporal organization of the project is presented, with tasks assigned to each year of the PhD programme, while at the same time two main technical stages, "final" and "far" rendezvous are introduced together with their scientific objectives. Being an experimental work, the envisioned implementation of the system on COTS computing platforms is introduced as well as the experiments planned to gather real imagery to validate the algorithms. Finally, possible fields of application of the project are discussed.

Introduction

Among several kind of fields, space is becoming one of the highest growing businesses in the world. With private companies paving the way, it is estimated that the number of satellites will increase of 2500 units every year for the next 10 years, with constellation making up the 83% of this number [1]. Indeed, several commercial services which were typically bounded to terrestrial infrastructure, are rapidly shifting to a space-based architecture (e.g. internet).

Due to relaying more and more on such space services, it becomes of the uttermost importance to ensure the continuity of operations of this plethora of objects operating on orbit.

Providing spacecraft with autonomous capabilities in terms of Guidance, Navigation and Control (GNC) has proven to bring many advantages [2]: the number of objects on orbit is increasing, hence their tracking for safety purposes is becoming difficult to perform if only ground structures are involved. By providing spacecraft with autonomous capabilities, the tracing load on the aforementioned structures can be relaxed while contextually ensuring the safety. Moreover, when distance between spacecraft is reduced, the latency of ground structures and human operator to close the GNC loop might be excessive and a collision could happen [3]. Developing and implementing autonomous capabilities could also pave the way to more complex space missions, either on close celestial bodies like the Moon and Mars, but also to implement On Orbit Servicing.

With this premises, the objective of the author's research project can be outlined: developing, testing and implementing a vision-based autonomous navigation system to perform a rendezvous with an uncooperative unprepared satellite on Commercial Off the Shelf computing platforms. Differently from other past projects, the absence of a detailed 3D model of the targeted spacecraft will be assumed.

Project outline

The project is divided in several stages, according to the three years of the PhD program. The first year will be dedicated to perform an extensive literature review to understand the actual state-of-the-art on the matter and to start design the prototypical detection and pose estimation algorithm. At the moment, the project is in this phase. The second year is dedicated to consolidate the

Aerospace Science and Engineering - III Aerospace PhD-Days   Materials Research Forum LLC
Materials Research Proceedings 42 (2024) 150-153   https://doi.org/10.21741/9781644903193-33

algorithm designed as well as to implement it on the target hardware/software platform, identified between the end of the first year and the beginning of the second. As it will be described in the following sections, real imagery gathered with sounding balloons will be utilized to validate the algorithm. Finally, during the last year the system will be integrated and an extensive campaign of hardware in the loop testing will be conducted to validate the results obtained. The last 5 months of the same year will be reserved for the writing of the final thesis.

Methodologies

Shifting to technical point of view, the project is divided in two parts: the first one involves developing a method to estimate the relative navigation parameters of a target (relative position and attitude) when the chaser is at a distance of less than 300 meters. This phase will be called "final rendezvous". The second stage instead is focused on estimate the same parameters but from a far higher distance, ideally of kilometres. This second phase will be called "far rendezvous". The difference between the two stages lies on the metrics that need to be estimated and the accuracy requested, hence the methods to be applied. Nevertheless, at the end of the project, the final system shall be capable of implementing both methodologies on a computing platform comparable to those already flying in the class of minisatellites.

The rendezvous distances are being adapted from International Rendezvous System Interoperability Standard [4].

Final rendezvous

Despite the name, this stage was chosen as the starting point for the project due to the relative larger availability of literature on the matter. The objective of this phase, as anticipated before, is to estimate the pose of a generic target using solely vision-based techniques at a distance comprised between 300 and 0 meters. By using the same standard cited before, the accuracy requirements are between 1° and 5° degrees for attitude and between 0.01 and 1 meter for range.

In the past few years, several methods to estimate the asset of an object have been investigated, either related to space or to automotive as well as to medical imaging, but objects in space presents several features that makes them difficult to analyse. On the matter, in [5] it is presented a review of the Kelvin Satellite Pose Estimation Challenge, aimed at pushing the state-of-the-art of the problem. From the results reported, two technical solutions seem applicable, if properly tuned, to our project: the separation between the localization phase and pose estimation phase of the target and the implementation of one or more Convolutional Neural Network (CNN).

Without going into the details of the CNN architectures implemented, both solutions outperformed the other methods in terms of accuracy. Nevertheless, adding another layer of computation (localizing the target and cropping consequently the frame) could increase the overall complexity of the systems. Paired with the hardware resources required for a highly performing CNN, implementing both solution on a relatively simple platform could become tough if proper measures are not taken. Such measures will be studied in the following stages of the PhD project. The aforementioned paper mentioned also another approach, detailed in [6], based on perceptual organization. Despite the method makes use of an a priori knowledge of the target, it could be a good candidate for our study due to lower computational costs if compared to CNN based approaches.

Far rendezvous

While the final part of the rendezvous process is being analysed and technical solutions explored, the initial part is not yet under intense examination for the reasons listed before. Nevertheless, considering the limits of using mostly passive sensors, some methods of identification and estimation could be adopted from the military world. In particular, an approach based on kinematic ranging [7] is deemed the best candidate for estimating the position of an object distant kilometres. This approach assumes the possibility to track the target and by performing a manoeuvre, thanks to a controlled variation of the angle between target and chaser, the position is triangulated with an acceptable accuracy. If observed carefully, the procedure proposed seems to imitate the estimations which can be obtained using stereo cameras.

Utilizing Stereo cameras is still considered but, due to the short distance achievable between the two sensors, it is not supposed to deliver significant results.

Implementation and planned experiments

Implementation on hardware platform

As anticipated in the introduction, the whole project aims at developing an autonomous navigation system to be implemented on a computing hardware comparable to those already flying. As a means of example, CubeSat is an appropriate candidate platform.

In simple terms, this means being able to perform the required estimation with limited computing power, if compared to that available while prototyping on a commercial laptop. The reasons to such design choice are multiple: the first one is that the power required by high performance computing platforms is not compatible with space applications, or at least not with those constrained by small solar panels and economic resources. Indeed, cost is another factor that must be taken into consideration: differently from the past, the space industry is moving to cheaper orbital platforms to widen the access to orbit and increase profits.

By implementing the system on simpler platform, within the price range of a generic Raspberry Pi 4 and/or a FPGA Intel Cyclone, hardware in the loop testing is possible, hence the validation of algorithms with greater fidelity. This requirement imposes to perform code optimization in order to deliver the desired performances within acceptable latency values. Hence the conversion of algorithms from high level programming (i.e. prototyped on Matlab) to an optimized language suitable for the real time operating system adopted (at this stage of the project is envisioned using Sel4 or ROS). Nevertheless, this could also be a chance to experiment with optimization techniques which could not be explored otherwise.

Planned experiments

Given the nature of the project, several campaigns of test are required to gather as much data as possible. Indeed, differences in performances were observed in [5] where vision systems were tasked to estimate the pose of an object both from synthetic and real images.

Considering the remote feasibility of conducting a "space test" campaign, sounding balloons were deemed proper candidates to simulate the imaging condition of open space. Indeed, these types of balloons reach an altitude of around 30-40 km, where air is mostly absent and at the same time lighting conditions are as harsh as those on higher orbits.

The flight test campaign is structured following a buildup approach with the first launch, estimated to be in mid-April, will be focused on validating the experimental set-up, schematized in figure 1. A note on the set-up is necessary: considering the difficulties on launching 2 independent balloons, being one the chaser and one the target, and acquire useful image data together with relative positions, it was decided to design a structure in which the target and the chaser are linked together. In particular, the action cameras (being the main sensor at this stage of the project) are integrated on the main structure just below the ballon and point to targets (basically cubes of predetermined dimensions) attached to a 1.5 meters beam. Targets have some freedom of movements and are provided with an asset recording device which will be used as source of truth for the pose estimation, which will be performed offline.

A similar test campaign is set to be happen at the end of May with the same set-up or with eventual modifications required after the first launch and starting from this, once the basic pose estimation algorithm is completed, a campaign with on-line pose estimation could be organized.

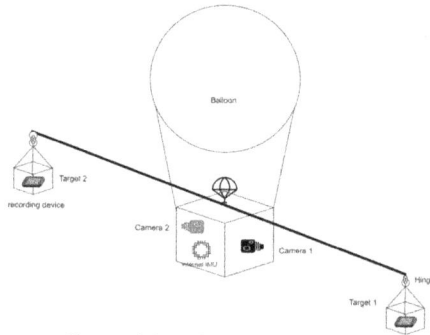

Figure 1 Sounding balloon set-up

Applications

Developing an autonomous vision-based navigation system can bring many advantages in different applications, the most obvious being On Orbit Servicing, active collision avoidance and automated rendezvous. On Orbit Servicing, enabled by automated rendezvous capabilities, could revolutionize the concept of orbital operations, both in terms of extending the useful life of precious spacecraft (similarly to what has been done with Hubble Space Telescope with the Shuttle program) but also when a controlled decommissioning of uncontrollable satellites is needed. Automating the approaching and the rendezvous phases becomes crucial when the aforementioned operations are performed away from Earth, where a human command is unpractical or at least not compatible with the latency required. On the side of long-distance operations, active collision avoidance systems could gain benefits from this project by scheduling more efficient manoeuvres thanks to a constant situation awareness of the surroundings of the interested spacecraft.

Reference

[1] D. Wet, "The Fast Mode," February 2023. [Online]. Available: https://www.thefastmode.com/expert-opinion/29970-satellite-trends-for-2023-and-beyond.

[2] A. Comellini, "Vision-based navigation for autonomous rendezvous with non cooperative targets," Université Fédérale Toulouse Midi-Pyrenées, Toulouse, 2021. https://doi.org/10.1109/IISA50023.2020.9284383

[3] B. Iannotta, "U.S. Satellite destroyed in space collision," Space.com, 12 February 2009. [Online]. Available: https://www.space.com/5542-satellite-destroyed-space-collision.html.

[4] W. H. Gerstenmaier, D. Parker, G. Leclerc and R. Shirama, *INTERNATIONAL RENDEZVOUS SYSTEM INTEROPERABILITY STANDARDS (ISRIS)*, 2019.

[5] M. Kisantal, S. Sharma, Y. H. Park, D. Izzo, M. Martens and S. D'Amico, "Satellite Pose Estimation Challenge: Dataset, Competition Design and Results," in *IEEE Transaction on aerospace and electronic systems*, 2020. https://doi.org/10.1109/TAES.2020.2989063

[6] S. D'Amico, M. Benn and J. L. Jorgensen, "Pose estimation of an uncooperative spacefraft from actual space imagery," *Journal of Space Science and Engineering*, vol. 2, no. 2, pp. 171-189, 2014. https://doi.org/10.1504/IJSPACESE.2014.060600

[7] P. Randall, C. Tucker and G. Page, "Kinematic ranging for IRST," *SPIE Acquisition, Tracking and Pointing*, vol. 1950, no. 7, 1993.

Aerospace Science and Engineering - IV Aerospace PhD-Days
Materials Research Proceedings 42 (2024) 154-157

Materials Research Forum LLC
https://doi.org/10.21741/9781644903193-34

Short and long-term reconstruction of in-orbit fragmentation events

Francesca Ottoboni[1,a] *, Andrea Muciaccia[1], Camilla Colombo[1]

[1]Department of Aerospace Science and Technology, Politecnico di Milano, Italy

[a]francesca.ottoboni@polimi.it

Keywords: Fragmentation Events, Debris Characterisation, Backward Propagation

Abstract. Over the past decade, the number of space debris has steadily increased. Consequently, the risk of collision between debris and active satellites has also increased, threatening the safety of space operations. Therefore, it is crucial to characterise fragments as soon as possible after their formation, to gather information about the fragmentation event which has generated them. In this context, the PUZZLE software has been developed at Politecnico di Milano to reconstruct past in-orbit breakups. This research aims at improving the current routine to obtain more accurate results and at optimising it for the analysis of fragmentations in the GEO and MEO regions.

Introduction

The exploitation of the space environment has massively increased over the last decade, resulting in a growing number of services relying on in-orbit satellites. In recent years, space debris have become a major concern for space agencies worldwide, as the majority of the population in space is represented by fragments which have been generated by in-orbit fragmentations such as collisions and explosions. Considering this scenario, the Inter-Agency Space Debris Coordination Committee has formulated mitigation guidelines concerning the disposal of satellites, yet the compliance to these measures is still too low for a sustainable use of space, hence the number of space debris is increasing. The consequence of this growth is a higher collision risk between satellites and fragments. Fragmentation events are difficult to predict, however the constant tracking of space objects allows to detect newly generated fragments. The analysis of the fragments upon formation is crucial to assess whether they have been generated in a new fragmentation event, allowing to characterise the breakup and reduce the risk it poses to other satellites.

Several methods have been developed to this aim, exploiting different features of the orbital motion of the fragments. Andrisan et al. [2] developed the Simulation of On-Orbit Fragmentation Tool (SOFT), to determine the epoch and the objects involved in a breakup using backward propagation and the position of the centre of mass.

Dimare et al. [3] characterise a fragmentation by defining an orbital similarity function between the orbital elements of the detected fragments. Several orbital similarity functions have been evaluated to correlate fragments with known orbits to parent(s), like it is usually done for asteroid families.

Frey et al. [4] developed a method to search for a fragmentation event in the long-term, i.e. in the order of years, exploiting a continuum approach for fragmentation modelling in LEO, based on the density of objects. The estimation of the epoch of the breakup is carried out by looking for the convergence of objects in inclination and right ascension of the ascending node (RAAN), which have been considered as robust features in LEO.

This work focuses on the PUZZLE software [1,5], which was developed at Politecnico di Milano with the purpose of detecting and analysing occurred fragmentations and the corresponding objects involved in them.

Aerospace Science and Engineering - III Aerospace PhD-Days
Materials Research Proceedings 42 (2024) 154-157

Materials Research Forum LLC
https://doi.org/10.21741/9781644903193-34

Methodology

The PUZZLE software analyses a set of unclassified Two-Line-Element (TLE) data to assess if a fragmentation has occurred in the recent past in Low Earth Orbit (LEO). The software has been developed for the reconstruction of fragmentation events both in the short-term (order of days) [5] and long-term (order of months to years) [1], however the methodologies implemented differ according to the time frame. The first step in the routine is the pre-filtering of the TLE data, which is performed in both cases. The aim of this step is to filter out outliers in TLEs, which could be due to errors in the initial orbit determination process [1]. For each set of TLEs associated to the same Satellite Catalogue Number, the filtering algorithm proposed by Lidtke et al. [6] is implemented.

A significant difference between the two versions of PUZZLE is the type of orbital elements used in the analysis. The short-term routine employs osculating orbital elements, which are propagated backwards with the Standard General Perturbations 4 (SGP4) propagator [7]. The long-term version instead uses mean Keplerian orbital elements, propagated with the PlanODyn propagator [8]. As opposed to SGP4, PlanODyn requires the Ballistic Coefficient (BC) for the propagation, therefore in the long-term PUZZLE routine the BC is estimated according to the method by Gondelach et al. [9] after the pre-filtering module, introducing a significant source of uncertainty.

The initial fragments are grouped into different families according to their orbital characteristics by means of pruning and clustering algorithms. Several methods have been implemented:

- A triple-loop filter, similar to the one proposed by Hoots et al. [10];
- An orbit inclination filter;
- An orbit RAAN filter;
- A Hierarchical Clustering Method (HCM) as in Zappala et al. [11].

The triple-loop filter consists in two geometrical filters and a time one. Its purpose is the identification of the fragments for which a close approach is possible while discarding the rest. The first filter is an apogee-perigee filter, to check if the geometry of the analysed pairs of objects is compatible. The second filter evaluates the MOID of the pairs of orbits. A threshold is set to reject the objects whose MOID is too high. The last filter generates angular windows around the MOID, converts them into time windows and searches for overlapping windows for each pair of objects. The triple-loop is implemented in both versions of PUZZLE, however the time filter is not used in the long-term investigation as it loses accuracy far from the event [1].

The inclination and RAAN filters are implemented only in the long-term version of the software. The inclination filter exploits the natural feature of LEO orbits for which inclination is little affected by perturbations. The RAAN filter is based on the boundedness of the RAAN for fragments and parent objects candidates near the event epoch [1].

The HCM is implemented only in the short-term routine. The algorithm is exploited to divide the objects into groups according to their orbital parameters at the fragmentation epoch. A similarity distance function is defined starting from the one proposed by Zappala et al. [11] and modifying it to account for all orbital parameters. The metric is evaluated for each couple of objects, grouping together the closest ones in an iterative approach until all the objects are assigned to a group and families have been formed.

After the backward propagation of the objects, the epoch estimation and parent identification of the fragmentation event is carried out. In the short-term method, the triple-loop is coupled with the propagator to detect clusters of objects in terms of position in time. Then, the HCM is employed to identify the fragments and the parents involved in the event. The long-term analysis instead exploits the RAAN filter to identify clusters of objects while backpropagating. A 2D histogram in sine and cosine of the RAAN allows to count the objects in each angular region and discard the objects which were not involved in the event. The parent identification process is carried out by

Aerospace Science and Engineering - III Aerospace PhD-Days
Materials Research Proceedings 42 (2024) 154-157

Materials Research Forum LLC
https://doi.org/10.21741/9781644903193-34

matching the candidates with a catalogue. The last step of the PUZZLE approach consists in simulating the fragmentation with the NASA standard breakup model [12,13], to estimate the number and the distribution of the generated fragments.

Limitations and future work

The PUZZLE software has been tested to reconstruct several known fragmentation cases, giving accurate results. However, some limitations exist. The algorithms implemented in the routine are sensitive to the size of the TLE set used as input, leading to significant computational times. The bottleneck is the triple-loop module, hence future work will be dedicated to its improvement from a computational point of view, investigating parallel computing strategies. Moreover, the algorithms will be revised to search for new possible pruning and clustering criteria and to improve the existing ones. For the long-term version, the revision will include a refinement in the computation of the BC which is a significant source of uncertainty.

A further step will be taken to harmonise and integrate the long-term and short-term versions of the software. The main idea is to include settings able to automatically switch from the long to the short-term search and vice versa according to the timeframe of the event and the input TLEs. In this way, the short-term investigation could also be used as a refinement of the long-term one [1]. In this framework, the PUZZLE approach will be improved in terms of automatisation of the selection of the settings for the automatic recognition of breakups. Appropriate values of the thresholds required by the algorithms will be investigated and validated through sensitivity analyses, such that the selected settings could be used for any fragmentation according to the orbital region.

Another limitation of the current versions of the software is that they have been designed to work only in LEO, as they exploit some natural features of this orbital region. Indeed, when trying to apply the same algorithm to fragmentation events at higher altitudes, the fragmentation identification fails. Therefore, the goal for future work is the extension of PUZZLE to other orbital regions (i.e., GEO and MEO) to allow the early analysis of detected fragments.

The last improvement of the software deals with the inaccuracy of the available data about space debris, which impacts the results achievable with PUZZLE. For this reason, future work will introduce uncertainty propagation in the algorithm by adding additional TLEs. Several uncertainty propagation methods will be considered, however the Gaussian Mixture Model and Unscented Transformation are expected to be the most suitable ones to properly represent the non-linearity of the problem. Preliminary analyses [14] have shown an improvement in the accuracy of identification of the event epoch and of the parent object(s) when including uncertainties, hence it is expected that the software will work also for cases that failed before, while reducing the sensitivity to input control parameters.

Conclusions

As the space debris population keeps growing, it is necessary to be able to track and characterise fragments originated in fragmentation events. In this work, the PUZZLE software for the detection and analysis of breakups has been presented. Two separate versions of PUZZLE can deal with fragmentation events occurred in the past in the short and long-term. Despite the good results obtained with the current routines applied to known fragmentation events, future work will be dedicated to the improvement of the implemented algorithms to guarantee more accurate results in a reduced computational time. Moreover, it will be crucial to extend the software to the GEO and MEO regions, to have a more global vision of the debris population.

Acknowledgments

This research received funding as part of the work developed for the project "Servizi inerenti alla realizzazione di un'infrastruttura HW e SW presso il CGS/Matera from the Italian Space Agency and from the European Space Agency contract 4000143257/23/D/SR "On-orbit breakup forensics".

References

[1] Muciaccia, A., Romano, M., & Colombo, C. (2021). Detection and characterisation of in-orbit fragmentations over short and long periods of time. Proceedings of the International Astronautical Congress, IAC, A6.

[2] Andrişan, R. L., Ioniţă, A. G., González, R. D., Ortiz, N. S., Caballero, F. P., & Krag, H. (2017). Fragmentation Event Model and Assessment Tool (FREMAT) Supporting on-Orbit Fragmentation Analysis. 7th European Conference on Space Debris, Darmstadt, Germany.

[3] Dimare, L., Cicalò, S., Rossi, A., Alessi, E. M., & Valsecchi, G. B. (2019). In-orbit fragmentation characterization and parent bodies identification by means of orbital distances. First International Orbital Debris Conference (IOC), Sugarland, Houston, TX, USA.

[4] Frey, S., Colombo, C., & Lemmens, S. (2018). Advancement of a density-based debris fragment model and application to on-orbit break-up localisation, 36th IADC plenary meeting, Jun. 5-8 2018, Tsukuba, Japan.

[5] Romano, M., Muciaccia, A., Trisolini, M., Di Lizia, P., Colombo, C., Di Cecco, A., & Salotti, L. (2021). Characterising in-orbit fragmentations with the PUZZLE software. 8th International Conference on Astrodynamics Tools and Techniques, Darmstadt, Germany.

[6] Lidtke, A., Gondelach, D. J., & Armellin, R. (2019). Optimising filtering of two-line element sets to increase re-entry prediction accuracy for GTO objects. Advances in Space Research, vol. 63, pp. 1289-1317. https://doi.org/10.1016/j.asr.2018.10.018

[7] Vallado, D. A., Crawford, P., Hujsak, R, et al. (2006). Revisiting Spacetrack Report #3. https://doi.org/10.2514/6.2006-6753

[8] Colombo, C. (2016). Planetary Orbital Dynamics (PlanODyn) suite for long term propagation in pertubed environment. Proceedings of the 6th International Conference on Astrodynamics Tools and Techniques (ICATT), Darmstadt, Germany.

[9] Gondelach, D. J., Armellin, R., & Lidtke, A. (2017). Ballistic Coefficient Estimation for Reentry Prediction of Rocket Bodies in Eccentric Orbits Based on TLE Data. Mathematical Problems in Engineering. https://doi.org/10.1155/2017/7309637

[10] Hoots, F. R., Crawford, L. L., & Roehrich, R. L. (1984). An analytic method to determine future close approaches between satellites. Celestial Mechanics, vol. 33, pp. 143-158. https://doi.org/10.1007/BF01234152

[11] Zappala, V., Cellino, A., Farinella, P., & Knezevic, Z. (1990). Asteroid families. I - Identification by hierarchical clustering and reliability assessment. The Astronomical Journal, vol. 100, pp. 2030-2046. https://doi.org/10.1086/115658

[12] Johnson, N. L., Krisko, P. H., Liou, J. C., & Anz-Meador, P. D. (2001). NASA's new breakup model of EVOLVE 4.0. Advances in Space Research, vol. 28, pp. 1377-1384. https://doi.org/10.1016/S0273-1177(01)00423-9

[13] Olivieri, L., Giacomuzzo, C., Lopresti, S. et al. (2023). Simulation of In-Space Fragmentation Events. Aerotec. Missili Spaz. https://doi.org/10.1007/s42496-023-00186-1

[14] Detomaso R. (2022). "Hybrid gaussian mixture model and unscented transformation algorithm for uncertainty propagation within the PUZZLE software", Politecnico di Milano, MSC in Space Engineering, Supervisor: Camilla Colombo, Co-supervisor: Andrea Muciaccia.

Aerospace Science and Engineering - IV Aerospace PhD-Days
Materials Research Proceedings 42 (2024) 158-161

Materials Research Forum LLC
https://doi.org/10.21741/9781644903193-35

Fuel-optimal low-thrust CAM with eclipse constraints

Eduardo Maria Polli[1,a] *, Juan Luis Gonzalo[2,b] and Camilla Colombo[3,c]

[1]Ph.D. student, Politecnico di Milano, Via La Masa 34, 20156 Milano, Italy

[2]Ph.D. Co-Supervisor, Assistant Professor, Politecnico di Milano, Via La Masa 34, 20156 Milano, Italy

[3]Ph.D. Supervisor, Associate Professor, Politecnico di Milano, Via La Masa 34, 20156 Milano, Italy

[a]eduardomaria.polli@polimi.it, [b]juanluis.gonzalo@polimi.it, [c]camilla.colombo@polimi.it

Keywords: Collision Avoidance Manoeuvre, Low Thrust, Optimal Control, Collision Probability, Space Traffic Management

Abstract. This works presents an optimal control direct method to design fuel-optimal low-thrust Collision Avoidance Manoeuvres (CAMs), imposing return to nominal conditions and including eclipse constraints. The methodology exploits the Hermite-Simpon integration scheme to impose the dynamics and the Squared Mahalanobis Distance (SMD) to ensure a collision probability lower than a prescribed value. The choice of using a direct method to solve the Optimal Control Problem (OCP) is justified by the inclusion of orbital perturbations.

Introduction

Satellite Collision Avoidance Manoeuvres (CAMs) play a crucial role in ensuring the safety and longevity of space missions within the increasingly crowded Low Earth Orbit region. As the number of operational satellites continues to rise, there is a growing emphasis on enhancing autonomy in both ground-based and space operations. Managing satellite functions, which are constrained by limited on-board resources, necessitates the deployment of lightweight algorithms that achieve a reasonable balance between efficiency and accuracy. Furthermore, this challenge is rendered more arduous by the increasing use of low-thrust propulsion systems, which necessitate control actions to operate over time intervals rather than relying on instantaneous impulsive manoeuvres.

Designing a CAM generally involves solving an Optimal Control Problem (OCP). The concept of optimality is typically measured in terms of maximising the miss distance, which represents the minimum separation between two approaching objects, or minimising the collision probability. The latter approach is more accurate because it accounts for uncertainties related to orbit determination for the objects of interest.

When dealing with low-thrust propulsion, achieving these optimal solutions often requires continuous, full-throttle thrust. However, energy is a limited resource onboard spacecraft, and minimising waste becomes a primary concern. Excluding natural perturbation exploitation [1], the OCP has to be reformulated as a fuel optimisation problem, ensuring either a minimum miss distance or a maximum probability of collision, also known as Accepted Collision Probability Level (ACPL).

State of the art CAM design

Numerous methods exist in the literature to address low-thrust CAM OCPs. One popular approach is based on the Sims-Flanagan transcription [2], where continuous low-thrust control is approximated as a series of impulsive manoeuvres. The continuous domain is discretised into small arcs such that an impulsive manoeuvre at the centre of the arc would provide the same Δv as if the

Aerospace Science and Engineering - III Aerospace PhD-Days
Materials Research Proceedings 42 (2024) 158-161

Materials Research Forum LLC
https://doi.org/10.21741/9781644903193-35

satellite was using low thrust for the whole arc. This discretisation enables the solution of the OCP with Nonlinear Programming (NLP) techniques.

More recently, Bombardelli and Hernando-Ayuso [3] and Gonzalo et al. [4] proposed two similar methods to determine analytically the optimal thrust direction of an impulsive manoeuvre, using Dromo and Keplerian elements, respectively. The main drawback of this formulation is that the model is based on the assumption of zero miss distance at the Close Approach (CA), i.e. the instant at which the separations between the two object is minimum.

A different approach is to solve the optimal control problem using an indirect method. In [5], De Vittori et al. propose an analytical solution of the energy optimal problem through the linearisation of the relative dynamics employing State Transition Matrices (STMs). This solution is then used as initial guess for the fuel optimal problem, which is solved as a Two-Points Boundary Value Problem that targets the desired SMD.

Convex techniques, such as Sequential Convex Programming (SCP), have also been applied to the problem. More recently, the use of AI decision-making tools is becoming increasingly popular. There are many operational aspects that most of the numerically efficient algorithms fail to consider. Particularly, these are the effect of uncertainty on the outcomes from the nominal thrust profile, returning to nominal conditions after the CAM is performed, the effects of orbital perturbations, and thrust constraints during eclipses. The first two problems are tackled separately in [6,7], respectively, while eclipses remain still untreated.

Motivation

The objective of this work is to develop a comprehensive framework for CAMs design, capable of including all relevant features. Notably, these are: fuel optimality, eclipses, perturbations, and return to nominal conditions. The proposed model is intended to serve as a robust foundation for ground-based computations and function as a benchmark for lightweight on-board algorithms.

Model

The proposed model is currently under development and the key design concepts are presented in this section. The dynamics are described in Keplerian elements using Gauss planetary equations, including perturbations, hereby expressed in the form:

$$\dot{y} = f(x, a_c, t)$$

Where $x = (a, e, i, \Omega, \omega, \theta)$ is the vector of Keplerian elements, a_c is the control acceleration, and t is time. Depending on the orbital regime, the perturbing acceleration vector can include the contributions due to geopotential (most notably $J2$, but other spherical harmonics can be considered in resonant regions), atmospheric drag and Solar Radiation Pressure (SRP). The general OCP is formulated as:

$$\begin{aligned} minimise \quad & J(y) \\ subject\ to \quad & h(y) = 0 \\ & g(y) \leq 0 \end{aligned}$$

Where $y = [x, a_c]$ is the optimisation variable composed of state x and control a_c, $J(y)$ is the objective function, or performance index, to be minimised, and $h(y)$ and $g(y)$ are, respectively, the equality and inequality constraints. In particular, the aim of the optimisation is to minimise fuel consumption, therefore the performance index is expressed as the integral of the acceleration magnitude over the entire time domain. The initial time t_0 and final time t_f are prescribed as user inputs. In particular, t_0 specifies the first available time to perform the manoeuvre, while t_f is the maximum time allowed to return to nominal conditions.

Aerospace Science and Engineering - III Aerospace PhD-Days   Materials Research Forum LLC
Materials Research Proceedings 42 (2024) 158-161   https://doi.org/10.21741/9781644903193-35

The equality constraints are used to impose the dynamics as defect constraints. Particularly, these are imposed exploiting a direct transcription and collocation method, using the Hermite-Simpson integration scheme. The defect constraint δ is the error between the analytical derivative computed at the mid-point of two nodes and the derivative estimated with Hermite-Simpson method at the same point:

$$\delta = \dot{x}_c - f(x_c, a_c, t_c)$$

Where the subscript c refers to the collocation mid-point.

This class of constraint is defined for every pair of adjacent nodes in order to impose the dynamics inside the optimisation procedure.

The second class of equality constraints is used to make the satellite return to nominal conditions after the CAM is performed. This can be achieved by imposing $x(t_f) = x(t_0)$, with the only exception of the true anomaly θ.

Inequality constraints are applied to enforce a minimum SMD at the encounter, and to ensure that the control action is performed outside the eclipse cone.

The SMD is measured on the b-plane and depends on the combined position uncertainty of the approaching objects. It is used in Chan's algorithm [8] to compute the respective collision probability. In particular, if the combined hard-sphere radius is also known, imposing a minimum SMD is equivalent to imposing a maximum ACPL. The respective constraint is defined as:

$$SMD(x, t_{CA}) > \overline{SMD}$$

Where t_{CA} is the time instant of the CA.

To avoid thrusting inside the eclipse cone, a constraint on the control a_c shall also be imposed, where $a_c = [a_r\ a_t\ a_n]$ is modelled in the RTN frame (radial, transversal, normal). Re-adapting the methodology presented in [9], it is possible to develop a function $e(x)$ that is negative when the satellite is inside the eclipse cone. The eclipse constraint could then be imposed as:

$$-e(x)(a_r^2 + a_t^2 + a_n^2) \leq 0$$

Results

At present, progress is confined to the creation of a direct method for evaluating fuel-optimal CAMs. This model, which is based on [1] and [3], uses impulsive manoeuvres to model the low-thrust profile. The optimality of each impulse is denoted by the eigenvalue associated to a specific manoeuvre. The fuel-optimal bang-bang control for low-thrust manoeuvres can be achieved through a one-degree-of-freedom optimisation, where the optimisation variable is the threshold eigenvalue that acts as switching function. Fuel optimal results of impulsive manoeuvres are shown in Fig. 1 for prescribed miss distance, impact parameter and collision probability.

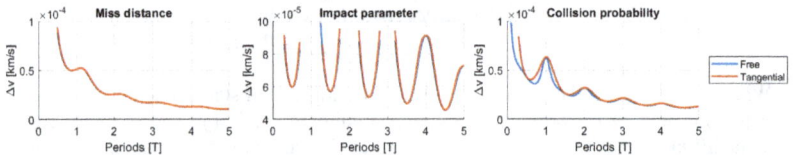

Figure 1 Required Δv to obtain a prescribed miss distance of 1 km, impact parameter of 250 m and collision probability of 10^{-6}. The red tangential curves refers to the impulsive being directed along tangential direction only.

Conclusion

The proposed model presents a novel solution for CAM design, where fuel-optimality, eclipse constraints and return to nominal conditions are included. The methodology exploits direct transcription to express the OPC as an NLP problem imposing both dynamics and operational constraints, in order to provide an efficient and complete decision support module for operators.

Acknowledgements

This research has received fundings as part of the project "Servizi inerenti alla realizzazione di un'infrastruttura HW e SW presso il CGS/Matera", from Italian Space Agency (ASI).

References

[1] Turco, F., Traub, C., Gaißer, S. et al. Analysis of Collision Avoidance Manoeuvres Using Aerodynamic Drag for the Flying Laptop Satellite. Aerotec. Missili Spaz. 103, 61–71 (2024). https://doi.org/10.1007/s42496-023-00183-4

[2] J.A. Sims, S.N. Flanagan, Preliminary Design of Low-Thrust Interplanetary Missions, Paper AAS 99-338 Astrodynamics Specialist Conference, Girdwood, Alaska (16-19 August 1999)

[3] C. Bombardelli, J. Hernando-Ayuso, Optimal Impulsive Collision Avoidance in Low Earth Orbit, Journal of Guidance, Control, and Dynamics, Vol. 38, No. 2, (2015) 217–225. https://doi.org/10.2514/1.G000742

[4] J.L. Gonzalo, C. Colombo, P. Di Lizia, Analytical Framework for Space Debris Collision Avoidance Maneuver Design, Journal of Guidance, Control, and Dynamics, Vol. 44, No. 3, (2021) 469–487. https://doi.org/10.2514/1.G005398

[5] A. De Vittori, M.F. Palermo, P. Di Lizia, R. Armellin, Low-Thrust Collision Avoidance Maneuver Optimization, Journal of Guidance, Control, and Dynamics, Vol. 45, No. 10, (2022), 1815–1829. https://doi.org/10.2514/1.G006630.

[6] M. Maestrini, A. De Vittori, J.L. Gonzalo, C. Colombo, P. Di Lizia, J. Míguez Arenas, M. Sanjurjo Rivo, A. Diez Martín, P. Gago Padreny, D. Escobar Antón, ELECTROCAM: Assessing the Effect of Low Thrust Uncertainties on Orbit Propagation. Paper 2nd ESA NEO and debris detection conference, Darmstadt, Germany (24-26 January 2023)

[7] A. De Vittori, G. Dani, P. Di Lizia, R. Armellin, Low-Thrust Collision Avoidance Design for LEO Missions With Return To Nominal Orbit, AAS/AIAA Space Flight Mechanics Meeting, Austin, Texas, (2023) 1–18.

[8] F.K. Chan, Spacecraft collision probability, Chapter 6. The Aerospace Corporation, El Segundo, California, United States of America, (2008) 13–97, https://doi.org/10.2514/4.989186

[9] Y. Wang, F. Topputo, Indirect optimization for low-thrust transfers with earth shadow eclipses, 31st AAS/AIAA Space Flight Mechanics Meeting, (1-4 February 2021) Virtual.

Aerospace Science and Engineering - IV Aerospace PhD-Days
Materials Research Proceedings 42 (2024) 162-168

Materials Research Forum LLC
https://doi.org/10.21741/9781644903193-36

Multilayer and multifield analysis of origami deployable structures

Tommaso Sironi[1,a] *, Maria Cinefra[1,b], Andrea Troise[1,c]

[1]Polytechnic of Bari, Italy

[a]t.sironi@phd.poliba.it, [b]maria.cinefra@poliba.it, [c]a.troise@phd.poliba.it

Keywords: Deployable Structure, Multilayer, Multi-Field, Carrera Unified Formulation, Reduced Models

Abstract. The research focuses on design of origami deployable structures for space applications. The aim is to acquire new and in-depth knowledge on modeling deployable structures with simple origami folding patterns to enhance the reliability of subsystem design requiring the use of such structures. Mastery of the subject will also enable the analysis of the compatibility of flexible thin structures with electronic components such as flexible rigid PCBs, solar cells, or antennas. The most used method of modelling for the analysis of flexible structures is based on bar-and-hinge models (reduced degree-of-freedom models) which, although accurately describe the macroscopic behavior, are unable to capture local behavior at critical points such as folds and interface points with bonded rigid-flexible elements. Consequently, a refined analysis of the system is carried out by integrating the finite element Carrera Unified Formulation (CUF) in the bar-and-hinge model, which allows to reduce the number of degrees of freedom with respect to classical Finite Element Method (FEM). Such formulation is necessary for two main reasons. Firstly, integrating this formulation with classical reduced degrees of freedom models will allow analyzing the local behavior of the deployable structure while still keeping computational costs low. Secondly, the CUF model is well suited for modeling thin multilayer structures, consisting of layers of very different materials, thanks to the possibility of selecting an arbitrary approximation order along the thickness that is independent of the order of the model adopted in the plane. Such a powerful tool will also allow analyzing the behavior at the interface between distinct layers subjected to both mechanical and coupled thermomechanical stresses, a critical condition in space applications. The final step of the modeling involves research on controlled deployment methods, deepened to ensure greater system reliability and morphing capabilities. The developed models will be experimentally validated through test campaigns aimed at verifying that the stress and strain states resulting from the analyses are comparable to those evaluated in experimental tests. These tests will include, among others, deployment tests of structures with origami patterns or classic folding patterns to evaluate opening stresses and cyclic thermal fatigue tests to evaluate thermomechanical stresses at the interface between distinct layers. The comparison will provide the numerical model with additional robustness and make it a tool capable of predicting complex behaviors otherwise investigable only through experimental tests.

Introduction

The space market has undergone significant transformations in the last two decades. The reduction in satellite launch costs and the pivotal role played by major private players have drastically accelerated the space race in both scientific research and service/business sectors. The future of the space economy appears increasingly promising, thanks to new technologies and growing interest from both agencies and private entities in accessing space.

Research and technological development have accompanied the sector's growth, consistently providing new solutions to the stringent requirements of spacecraft design and reducing production and launch costs.

Aerospace Science and Engineering - III Aerospace PhD-Days   Materials Research Forum LLC
Materials Research Proceedings 42 (2024) 162-168   https://doi.org/10.21741/9781644903193-36

In this context, deployable structures and Gossamers play a fundamental role in the design of both small and large satellites. These structures are designed to launch objects with significant volumes and surfaces while remaining compatible with the compactness requirements imposed by launchers. Given the numerous subsystems whose performance depends on their surface extension, these structures find applications in various fields, from power generation using solar panels to signal transmission with antennas, thermal control with thermal shields and radiative panels, and even propulsion and space based solar power generation for ground based business [1].

Although deployable structures are already employed in satellites and spacecraft and are of great technological and market interest, their modelling approach remains a topic of research. Specifically, for extensively large and origami-patterned thin structures, numerical models explaining their behaviour during operational life and deployment are challenging to represent using finite elements, making the computational calculation process cumbersome and complex. One solution to this issue is provided by reduced order models, which effectively represent the overall behavior of these structures but fall short in capturing local information. In the modeling of origami, the more widely used reduced order model is the bar-and-hinge, consisting of a simplification of the kinematics of foldable thin structure justified by the fact that the big-scale deformations are imposed by the geometry of the folds. The nonlinear elastic formulation for a general bar-and-hinge model introduced by Ke Liu and Glaucio H. Paulino [2] stands as one of the most advanced reduced models.

Despite the accuracy in modeling the global behavior, a finer discretization of the structure is necessary to understand its local behaviors. A finite element shell model seems to be suitable for this purpose, but it presents two significant limitations. Firstly, a shell element models the structure through its thickness as a homogeneous entity, losing information about behavior across the thickness. Secondly, the computational cost of performing nonlinear dynamic analysis of the deployment of rigid-flexible origami structures is significant, especially if multiple shell elements are to be coupled to characterize the behavior of distinct layers across the thickness. There are few studies in the literature that employ finite element models for this purpose, with a clear preference towards reduced element analysis.

Similar issues arise when conducting a multi-field analysis on the structure under consideration, such as thermal-structural coupling induced by thermomechanical stresses, a very common condition in the space field. Uncoupled thermoelasticity models (static, quasi-static, dynamic) are able to model the relationship between stresses/strains and temperature but are not suitable for conditions of high-speed thermomechanical loads. Coupled thermoelasticity models based, such as the one proposed by Green-Lindsay, come into play, allowing to capture the real physical behavior of the component through simulations of the interaction between the mechanical behavior of the elastic body and its temperature where the temporal derivatives of deformations appear in the heat equations. While these equations provide a significant advantage in terms of result accuracy, the coupled thermoelasticity problem entails very high computational costs, and finite element models become necessary. Such models can be further complex in the case of structures with real geometries and very complex boundary conditions, as well as composed of metamaterials, increasingly common in the aerospace field and difficult to model.

Bar-and-Hinge Model

The bar-and-hinge model is based on the principle of stationary potential energy and allows for the development of a nonlinear model in geometry and material characteristics for the analysis of large deformations in origami structures. The panel is discretized into a series of bar and hinge elements typically located at the folds of the pattern or the diagonal of the flexible faces. The discretization just described allows capturing the three fundamental behaviours of origami deformation: stretching, crease folding, and panel bending.

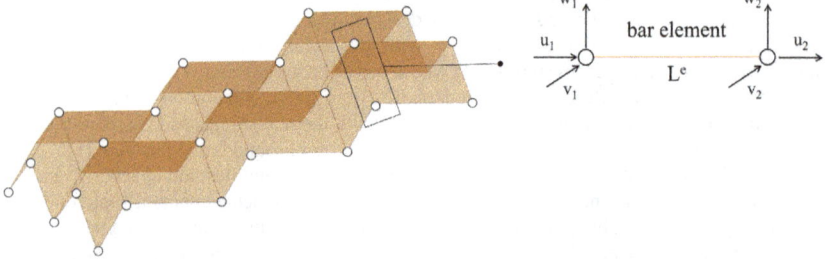

Figure 1: Miura-Ori bar-and-hinge model

In the static case analysis, the potential energy of the system is expressed as:

$$\Pi = U_{bar} + U_{spr} + V_{ext} \tag{1}$$

where U_{bar} is the strain energy of the bar elements, U_{spr} is the energy stored in the springs modelling the folds and the out of plane bending behaviour of the structure, while V_{ext} is the external work. In the dynamic case, the term E_k is introduced in the summation giving the potential energy and represents the kinetic energy of the system:

$$\Pi = U_{bar} + U_{spr} + V_{ext} + E_k \tag{2}$$

In the same way as for the potential energy, the tangent stiffness matrix can be decomposed into two contributions as follows:

$$\boldsymbol{K}(\boldsymbol{u}) = \boldsymbol{K}_{bar}(\boldsymbol{u}) + \boldsymbol{K}_{spr}(\boldsymbol{u}) \tag{3}$$

where $\mathbf{K}_{bar}$ is the stiffness associated with the bar elements while $\mathbf{K}_{spr}$ is the stiffness of the rotational springs modelling even the folding lines or the bending diagonal of the panel.
Each term can in turn be expanded in several matrices:

$$\boldsymbol{K}_{bar}^{(e)} = \boldsymbol{K}_E^{(e)} + \boldsymbol{K}_1^{(e)} + \boldsymbol{K}_2^{(e)} + \boldsymbol{K}_G^{(e)} \tag{4}$$

$$\boldsymbol{K}_{spr}^{(r)}(\boldsymbol{u}) = \widetilde{\boldsymbol{K}}_{spr}^{(r)}(\boldsymbol{x}) = k\frac{d\theta}{dx^{(r)}} \otimes k\frac{d\theta}{dx^{(r)}} + M\frac{d^2\theta}{d(x^{(r)})^2} \tag{5}$$

As regards the bars, $\mathbf{K}_E^{(e)}$ is the linear stiffness matrix, $\mathbf{K}_G^{(e)}$ is the geometric stiffness matrix and $(\mathbf{K}_1^{(e)} + \mathbf{K}_2^{(e)})$ form the initial displacement matrix. Concerning spring elements, θ is the dihedral angle of the rotational spring, $\mathbf{x}^{(r)}$ is the vector of nodal coordinates, while k is the tangent rotational stiffness of the element.
Bar and hinge elements can be modelled by different constitutive relations.
Odgen model can be implemented for the analysis of bar elements, according to the following expression:

$$W(E) = \widetilde{W}(\lambda_1, \lambda_2, \lambda_3) = \sum_{j=1}^{N} \frac{\mu_j}{\alpha_j}(\lambda_1^{\alpha_j} + \lambda_2^{\alpha_j} + \lambda_3^{\alpha_j} - 3) \tag{6}$$

Aerospace Science and Engineering - III Aerospace PhD-Days
Materials Research Proceedings 42 (2024) 162-168

Materials Research Forum LLC
https://doi.org/10.21741/9781644903193-36

Where λ_i denotes the principal stretches and N, α and μ are the material properties.

On the other hand, rotational hinges for origami structural analysis are usually modelled as linear elastic springs. The momentum generated in response to the external loads is as follows:

$$M = L^{(r)}k(\theta - \theta_0) \tag{7}$$

where k is the rotational stiffness modulus per unit length along the axis and θ_0 is the neutral angle at which the spring is in a stress-free condition. This relation can be generalized to implement the behaviour of nonlinear springs by modelling the stiffness as constant throughout most of its rotation range, while it reaches high values of stiffness in correspondence of a fully packed configuration of the origami.

The resolution of such models relies on numerical computation algorithms such as the Newton-Raphson method. The Merlin model for origami modeling employs a modified generalized displacement control algorithm directly derived from the arc length method.

Carrera Unified Formulation

The solution to the numerical computation problems highlighted in the introductory chapter is provided by the Carrera Unified Formulation (CUF). The choice to integrate the bar-and-hinge model with the finite element CUF is driven by the need to employ a local model for the detailed description of the behavior of deployable structures in space applications without the computational cost hindering analysis in reduced time frames. This formulation allows for deriving the governing equations in a compact way, it can decouple the level of accuracy in modeling along the thickness and in the plane of the thin structure, and it provides accurate solutions with a low number of Degrees Of Freedom (DOFs), without the need to resort to finite 3D elements of higher or lower order.

Below is the formulation of the CUF model for a thin plate. Let's consider a generic plate structure described in a Cartesian coordinate system. Let's consider the mid-plane of the plate lying on the xy-plane, while the thickness of the plate extends along z. The displacement field of a two-dimensional model in the CUF framework is described as a generic expansion of the generalized displacements (in the case of displacement-based theories) by arbitrary functions of the cross-section coordinates:

$$\mathbf{u}(x, y, z) = F_\tau(z)\mathbf{u}_\tau(x, y) \qquad \tau = 1, ..., M \tag{8}$$

where $u = \{u_x, u_y, u_z\}$ is the vector of 3D displacements and $u_\tau = \{u_{x\tau}, u_{y\tau}, u_{z\tau}\}$ is the vector of general displacements, M is the number of terms in the expansion, τ denotes a summation and the functions $F_\tau(z)$ define the approximation function along the thickness. The generalized displacements are function of the mid-plane coordinates and the expansion is conducted in the thickness direction z.

The same formulation can be applied to thermomechanical and electromechanical problems for the multifield analysis of structures such as piezoelectric ones.

The main advantage of CUF is that it allows to write the governing equations and the related finite element arrays in a compact and unified manner, which is formally an invariant with respect to the F_τ functions

In the case of 2D models, the discretization of generalized displacements on the mid-surface of the plate is made by means of the finite element method or other numerical methods.

Generalized Theory of Thermoelasticity

The CUF model can be employed in the thermomechanical analysis of components subjected to combined structural and thermal stresses [3]. The formulations of coupled thermoelasticity that

Aerospace Science and Engineering - III Aerospace PhD-Days   Materials Research Forum LLC
Materials Research Proceedings 42 (2024) 162-168   https://doi.org/10.21741/9781644903193-36

best represent the stress state are those belonging to the category of the Generalized Theory of Thermoelasticity.

According to Green-Lindsay (GL) and Lord-Shulman (LS) theories, the equation of motion of a 3D elastic body in physical coordinates (x, y, z) can be expressed in terms of displacement components as:

$$(C_{ijkl}u_{k,l})_j - (\beta_{ij}T)_j - (t_1\beta_{ij}\dot{T})_j + X_i = \rho_i\ddot{u} + \xi\dot{u}_i \tag{12}$$

Where X_i denotes the body forces per unit volume, ρ is the mass density, ζ is the damping coefficient of the material, u indicates the displacement components vector, C_{ijkl} is a fourth order tensor containing all the elastic coefficients of a general nonhomogeneous anisotropic material, β represents the second order tensor of thermoelastic moduli, T is the temperature change with respect to a reference temperature T_0 and t is time.

On the other hand, the energy equation can be expressed in terms of temperature and displacement fields as:

$$\rho c(t_0 + t_2)\ddot{T} + \rho c\dot{T} - 2\tilde{c}_i\dot{T}_i - (k_{ij}T_j)_i + t_0T_0\beta_{ij}\ddot{u}_{ij} + T_0\beta_{ij}\dot{u}_{ij} = R + t_0\dot{R} \tag{13}$$

Where k_{ij} is the thermal conductivity tensor, t_0 is the relaxation time associated with LS theory while $\tilde{c}$ is a relaxation time associated with GL theory, c is the specific heat.

The equations just reported constitute the governing system of equations for the generalized coupled thermoelasticity problems.

By applying a finite element formulation through Galerkin approach in a 3D domain, the associated week formulation containing all the possible boundary conditions of the problem is the following:

$$\sigma = C\epsilon - \beta(T + t_1\dot{T}) \tag{14}$$

The computational effort in solving the equation above is quite demanding because of the high number of degrees of freedom. To reduce the computational cost of such a problem without loosing in accuracy, refined 2D models can be implemented in the framework of the CUF presented in the previous paragraph.

In the final system of equation coming from the approximation with CUF approach, the stiffness matrix assumes the following expression:

$$K = \begin{bmatrix} K_{UU}^{lm\tau s} & K_{U\Theta}^{lm\tau s} \\ 0 & K_{\Theta\Theta}^{lm\tau s} \end{bmatrix} \tag{15}$$

The term K is the thermo-mechanical coupling term that models the mechanical stress induced by a thermal variation. The aim of the initial stages of the research path will be to implement the thermo-mechanical coupling term induced by a mechanical stress.

These coupling terms, combined with the capabilities of the CUF to model multilayer structures, allow for a thorough understanding of the behavior at the interface between two layers of a laminate or the junction between different materials in general [4], thus enabling the prediction of debonding or delamination.

Aerospace Science and Engineering - III Aerospace PhD-Days   Materials Research Forum LLC
Materials Research Proceedings 42 (2024) 162-168   https://doi.org/10.21741/9781644903193-36

Preliminary results

The case study considered for analysis is SolarCube, a flexible origami-inspired solar panel designed by Astradyne. The panel lends itself well to the type of analysis described due to the flexible nature of the substrate on which the rigid-flexible electronics are assembled. Moreover, the environment in which it operates is subject to significant temperature gradients, which can induce undesirable phenomena such as excessive relaxation of the substrate or delamination of material layers.

The deployment dynamics will be analyzed using the bar-and-hinge model integrated with CUF and will be compared with results obtained through commercial software such as Ansys and the bar-and-hinge reduced model alone. Figure 2 shows the deployment sequence of the origami with Merlin 2 software (bar-and-hinge).

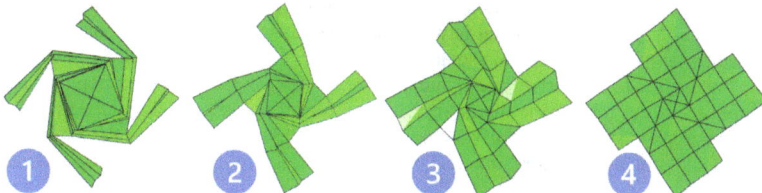

Figure 2: SolarCube's deployment phase. Figure reference: [7]

The structure is discretized with 5 nodes and 8 bars for each face of the origami. As for the material properties assigned, these are partly derived from literature data and partly obtained from mechanical characterization tests conducted in the laboratory. Finally, the deployment dynamics are induced by a displacement constraint at the vertices of the origami. The result is the displacement-force curve in Figure 3, showing a slow and gradual increase in the force required to open the origami and a sudden peak of force corresponding to the deployed configuration.

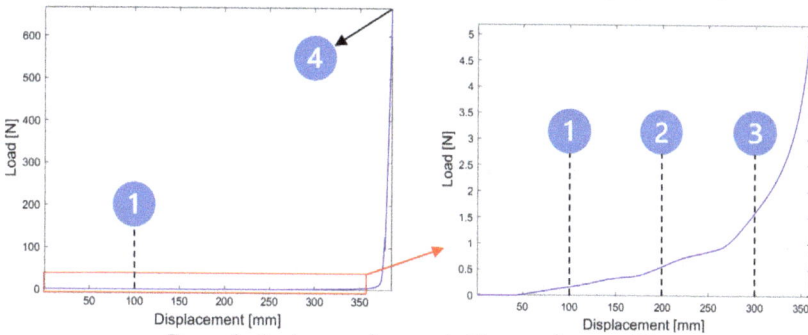

Figure 3: Deployment force path. Figure reference: [7]

The integrated bar-and-hinge and CUF model aims to capture the detail of the thermo-mechanical behavior of the panel at critical local points, potentially subject to delamination or concentrated stress.

Innovation and significant results

The first part of the research will focus on the thermo-mechanical numerical modeling of deployable metasurfaces for space applications. This modeling will include the two thermo-mechanical coupling terms induced by mechanical and thermal stresses, as well as the nonlinearities of the materials involved, considering the variation of their properties with temperature and fatigue conditions. Such analysis is essential in an environment like space, subject to a large number of thermal cycles over a wide range of temperatures. Further mastery of the results will be acquired by validating the model with experimental tests based on validation processes sanctioned by the ESA (European Space Agency) in the ECSS (European Cooperation for Space Standardization).

Subsequently, finite element models for the dynamics of the deployment of the flexible structure will be carried. The coupling of the CUF with the bar-and-hinge model will allow, unlike common reduced models, to analyze the local behavior of the panel and predict stress concentration points that may cause damage to the panel or delamination between the layers of metamaterial. The CUF and bar-and-hinge model are aimed to preserve the low computational cost of reduced models, while incorporating localized analysis at pre-selected points of interest within the simulation. Moreover, the model is well suited for thermo-electro-mechanical modelling of active deployment systems involving smart material such as shape memory alloys, paving the way for the analysis of innovative deployment systems for origami-inspired deployable space structures. As in the case of the thermo-mechanical model, experimental validation tests of the numerical model will follow to certify the quality of the work done and refine the model to obtain a powerful and effective tool for numerical modeling of structures in both academic and industrial fields.

References

[1] Erica Rodgers, Ellen Gertsen, Jordan Sotudeh, Carie Mullins, Amanda Hernandez, Hahn Le, Phil Smith, and Nikolai Joseph, 2024, Space Based Solar Power, NASA, Report ID 20230018600

[2] Liu K, Paulino GH. 2017 Nonlinear mechanics of non-rigid origami: an efficient computational approach.Proc. R. Soc. A 473: 20170348. http://dx.doi.org/10.1098/rspa.2017.0348

[3] A. Entezari, M. Filippi & E. Carrera (2017): Unified finite element approach for generalized coupled thermoelastic analysis of 3D beam-type structures, part 1: Equations and formulation, Journal of Thermal Stresses. https://doi.org/10.1080/01495739.2017.1336740

[4] Vincent Voet, Frederik Van Loock, Christophe De Fruytier, Aude Simar, Thomas Pardoen, Machine learning aided modelling of thermomechanical fatigue of solder joints in electronic component assemblies, International Journal of Fatigue, Volume 167, Part A, 2023, 107298, ISSN 0142-1123. https://doi.org/10.1016/j.ijfatigue.2022.107298.

[5] Cinefra, M., Rubino, A. Assessment of New Adaptive Finite Elements Based on Carrera Unified Formulation for Meshes with Arbitrary Polygons. Aerotec. Missili Spaz. 102, 279–292 (2023). https://doi.org/10.1007/s42496-023-00165-6

[6] Scano, D., Carrera, E. & Petrolo, M. Use of the 3D Equilibrium Equations in the Free-Edge Analyses for Laminated Structures with the Variable Kinematics Approach. Aerotec. Missili Spaz. (2023). https://doi.org/10.1007/s42496-023-00177-2

[7] Troise, Andrea. (2023). Reduced-order modelling of the deployment of a modified flasher origami for aerospace applications. 547-552. 10.21741/9781644902813-120.

Aerospace Science and Engineering - IV Aerospace PhD-Days   Materials Research Forum LLC
Materials Research Proceedings 42 (2024) 169-172   https://doi.org/10.21741/9781644903193-37

Towards a greener cosmos: Tool for integrating life cycle costs and sustainability for future space systems development

Sai Tarun Prabhu Bandemegala[1,a]

[1]Corso duca degli Abruzzi 24, 10129, Italy

[a]sai.bandemegala@polito.it

Keywords: Life Cycle Costs (LCC), Space Systems, Sustainability, Life Support Systems (LSS), Technology Readiness Level (TRL), Collaborative, Risk-Mitigation, Scalability

Abstract. As humanity ventures further into space exploration, the imperative for sustainability and cost-effectiveness intensifies. Accurate cost estimation is pivotal for project viability, funding, and stakeholder management. However, existing tools often fall short in integrating comprehensive data and sustainability considerations. This paper proposes the Sustainable Exploration Resource Toolkit (SERT), which integrates life cycle costs (LCC) analysis, sustainability principles, and technology evaluation to address these challenges. SERT aims to provide decision-makers with holistic insights into the financial implications of space missions, identifying cost-effective solutions for sustainable bioregenerative systems. By fostering cross-industry collaboration and complexity reduction strategies, SERT strives to revolutionize cost estimation for future space exploration endeavors, ensuring economic viability and environmental responsibility.

Introduction

In the vast expanse of space exploration, the quest for sustainability and cost-effectiveness has become increasingly paramount. As humanity sets its sights on long-duration space missions and interplanetary colonization, the need for innovative solutions that balance affordability with environmental responsibility has never been more critical. Accurate cost estimation is essential for ensuring project feasibility, securing funding, and managing stakeholder expectations. By understanding the full scope of financial requirements, space agencies and organizations can make informed decisions regarding mission planning, resource allocation, and risk management. Moreover, transparent and comprehensive cost estimation enables policymakers, taxpayers, and the public to assess the value and benefits of space exploration initiatives. In an era of budget constraints and competing priorities, having a realistic understanding of project costs is critical for maintaining public trust and support for space exploration endeavors. Additionally, accurate cost estimation helps mitigate the risk of budget overruns and delays, enabling projects to stay on schedule and within budget constraints. In response to these challenges, a paradigm shift that emphasizes the integration of life cycle costs (LCC) analysis [1] and sustainability considerations to pave the way towards a greener cosmos becomes predominant. Central to this approach is the development of methodologies and tools that support space systems developers and operators in selecting the most affordable and sustainable solutions for providing vital resources in space exploration missions.

At the core of this proposal lies the incorporation of life cycle costs (LCC) analysis, a comprehensive framework for assessing the total costs associated with each phase of space system development and operation. This analysis encompasses not only initial research and development costs but also manufacturing, assembly, and operational costs, providing decision-makers with a holistic view of the financial implications of space missions. By considering the entire life cycle of space systems, from inception to retirement, LCC analysis enables informed decision-making and the identification of cost-saving opportunities that may arise over time [2]. In parallel,

Aerospace Science and Engineering - III Aerospace PhD-Days   Materials Research Forum LLC
Materials Research Proceedings 42 (2024) 169-172   https://doi.org/10.21741/9781644903193-37

sustainability considerations play a crucial role in shaping the design and implementation of future space systems. These considerations encompass a broad range of factors, including resource utilization, waste management, and environmental impacts, with the overarching goal of minimizing ecological footprints and promoting eco-friendly practices in space exploration [3]. By integrating sustainability principles into space system development costs, the SERT tool can ensure that space missions align with long-term sustainability goals, fostering responsible stewardship of celestial resources for all the stakeholders involved.

Numerous tools exist to facilitate LCCA for space systems and they are not without limitations. One of the primary challenges faced by LCCA tools is the integration of comprehensive and accurate data throughout the life cycle of space systems. Often, these tools rely on historical data or assumptions due to the lack of real-time or predictive data sources. Factors such as evolving technology, regulatory changes, and unforeseen events can significantly impact cost estimations. Thus, the inability to incorporate dynamic data sources hampers the accuracy of LCCA. While ALiSSE, the Advanced Life Support Systems for Space Exploration tool developed by the European Space Agency (ESA), offers valuable support for designing and evaluating life support systems for space missions, it also presents several drawbacks. These include its complexity, which may require specialized knowledge to use effectively, and potential limitations in simulating certain aspects of life support systems. Additionally, ALiSSE's resource-intensive nature, both in terms of computational requirements and time, could pose challenges for users with limited resources or tight project timelines. The accuracy and reliability of ALiSSE's predictions may also depend on the quality of input data and assumptions, necessitating validation against real-world data. Furthermore, the user interface may not be user-friendly for all users, impacting usability and accessibility.

On the other hand, the NASA Equivalent System Mass (ESM) is a metric used by NASA to evaluate the mass of various life support systems and consumables required for space missions. ESM quantifies the total mass of a system or set of components needed to provide a certain level of support for astronauts over a specified duration. It encompasses factors such as food, water, air, waste management, and other consumables necessary to sustain human life in space. By calculating ESM, NASA can compare different life support system designs and evaluate their efficiency and feasibility for long-duration missions. However, ESM also has several drawbacks. One limitation is its focus solely on mass, which may not fully capture the complexity and nuances of life support system design. For example, ESM does not account for factors such as volume, power requirements, or the reliability of system components, which are also critical considerations for space missions. Additionally, ESM calculations may oversimplify the complexities of human physiology and behavior, leading to inaccuracies in estimating resource requirements. Furthermore, ESM assessments may not adequately address the integration of life support systems with other spacecraft systems or the dynamic nature of mission scenarios.

Many of these tools focus primarily on direct costs associated with the design, development, and operation of space systems. However, they overlook indirect costs, such as maintenance, upgrades, and disposal, which are integral parts of the life cycle as well. These tools also often struggle to incorporate risk and uncertainty factors adequately. While sensitivity analysis and Monte Carlo simulations can offer some insights, they do not capture the full spectrum of potential risks associated with space missions, leading to overly optimistic or pessimistic cost estimates. Moreover, many existing LCCA tools are designed for specific types of space missions or technologies, lacking scalability and flexibility to adapt to diverse project requirements. As a result, users find it challenging to customize the analysis framework according to their unique needs, limiting the applicability of these tools across different space programs and initiatives. Furthermore, factors such as evolving technology, regulatory changes, and unforeseen events can significantly impact cost estimations while often, the existing tools rely on historical data or assumptions due to the lack of real-time or predictive data sources.

Methodology and Expected Results

This project was proposed to address this critical need for comprehensive analysis of life cycle costs throughout the various phases of life support system development, which directly led to the development of the first iteration of Sustainable Exploration Resource Toolkit (SERT) tool that addresses key aspects essential for the success of long-duration space missions. The SERT tool proposed aims to address these issues by assessing existing and commercial technologies relevant to bioregenerative systems by gathering information on available technologies, their capabilities, performance metrics, and limitations within the tool system. In addition, it will also incorporate data from performance testing and validation studies derived from laboratory testing, ground-based simulations, or field trials to assess the functionality, reliability, and resilience of technologies in microgravity, radiation, and extreme temperature environments. Through this integration of comprehensive and accurate data throughout the life cycle of space systems, SERT can provide cost estimates that will be close to actual costs of the missions over time.

In addition, to mitigate the development risks associated with sustainable and cost-effective bioregenerative systems, SERT also incorporates the evaluation of existing and commercial technologies. By leveraging existing expertise and infrastructure, space agencies can reduce complexity in system components and processes, optimize operational efficiency, and minimize costs. Collaborations with terrestrial agriculture research centers offer invaluable opportunities to tap into the wealth of knowledge and experience in agricultural science, enabling the adaptation of proven technologies and methodologies for space applications. Moreover, the feasibility of utilizing existing technologies and collaborations with terrestrial agriculture research centers serve to optimize operational efficiency and minimize costs. By harnessing existing expertise and infrastructure, space missions can accelerate technology development, reduce research and development costs, and enhance the reliability and performance of bioregenerative systems. Since interdisciplinary collaborations have become the cornerstone of innovation in today's interconnected world by bridging the gap between industries, these partnerships have fueled progress, driven efficiency improvements, and unlocked new opportunities for growth and development. In that regard, healthcare has witnessed a revolution with the collaboration between medical professionals and 3D printing companies, leading to the creation of custom-made implants that reduce surgery time, improve outcomes, and save costs. Retailers partnering with tech companies utilize big data analytics to optimize supply chains, resulting in streamlined logistics, reduced inventory, and improved customer satisfaction. The aerospace industry's expertise in lightweight materials, transferred to automotive manufacturing, has yielded fuel-efficient vehicles that meet emission regulations. Therefore, integrating expertise from terrestrial agriculture research centers and utilizing existing technologies allows space missions to optimize operational efficiency and maximize yields, while minimizing resource consumption and waste generation. This ensures the long-term sustainability of space habitats while supporting the nutritional needs of astronauts during extended missions. Therefore, this SERT tool will also feature options to integrate the current, as well as up-coming technologies in the form of their Technology Readiness Levels (TRLs) to give an estimate of costs involved at various step-intervals.

Conclusion

In conclusion, the proposed SERT tool aims to underpin the development of space systems that are not only robust and economically viable but also environmentally responsible. By integrating life cycle costs (LCC) analysis, sustainability considerations, and complexity reduction strategies, SERT strives to address the multifaceted challenges inherent in space exploration. Through the tools' extensive analysis, decision-makers gain a comprehensive understanding of the total costs

Aerospace Science and Engineering - III Aerospace PhD-Days   Materials Research Forum LLC
Materials Research Proceedings 42 (2024) 169-172   https://doi.org/10.21741/9781644903193-37

associated with each phase of system development and operation, enabling informed decision-making and resource allocation instead of being limited by ESM or overstimulated by ALiSSE tools. Moreover, by systematically evaluating the feasibility of utilizing existing and commercial technologies, space agencies and organizations can identify low-risk, cost-effective solutions for developing sustainable and resilient bioregenerative systems for space missions. Thereby, these sustainability considerations ensure that space systems align with long-term environmental goals, minimizing ecological footprints and promoting eco-friendly practices. Additionally, complexity reduction strategies mitigate development risks and optimize operational efficiency, further enhancing the economic viability of space missions. Furthermore, by fostering cross-industry collaboration, businesses can unlock a wealth of potential for cost savings and efficiency gains. This collaborative approach not only strengthens individual companies but also propels technological advancements that benefit entire sectors and, ultimately, society as a whole. Implementing these methodologies into the tool holds the potential to revolutionize future space exploration missions, particularly in meeting the sustainable exploration needs of astronauts during extended missions. By deriving upon innovation, efficiency, and environmental responsibility, SERT paves the way for a new era of estimating costs for space exploration missions that is both technologically advanced and environmentally sustainable.

References:

[1] Hauschild, M. Z., Olsen, S. I., & Rosenbaum, R. K. (Eds.). (2018). Life cycle assessment. SpringerLink. https://link.springer.com/book/10.1007/978-3-319-56475-3

[2] Seedhouse, E. (2020). Life Support Systems for Humans in Space. In Springer eBooks. https://doi.org/10.1007/978-3-030-52859-1

[3] Dominoni, Annalisa. "Design for Sustainability in Space: Future Nature." SpringerBriefs in Applied Sciences and Technology, 1 Jan. 2024, pp. 25–48, https://doi.org/10.1007/978-3-031-56004-0_2. Accessed 2 Apr. 2024.

Aerospace Science and Engineering - IV Aerospace PhD-Days
Materials Research Proceedings 42 (2024) 173-177

Materials Research Forum LLC
https://doi.org/10.21741/9781644903193-38

A roadmap for the implementation of augmented reality solutions for airport control towers in an operative environment

Tommaso Fadda*

Department of Industrial Engineering, Università di Bologna, 47121, Forlì, Italy

tommaso.fadda2@unibo.it

Keywords: Extended Reality, ADS-B, Airport Control Tower, Advanced Air Mobility, AAM, Digital Towers

Abstract. This paper describes how augmented reality has been recently exploited to define novel concepts for helping control tower operators in the aerodrome traffic zone. Starting from solutions developed within the SESAR research framework through real-time simulations, the paper describes the work done so far in transferring the developed solutions to an operative scenario (i.e. an actual control tower), exploiting ADS-B data and a HoloLens2™ head-mounted device. Finally, a roadmap covering the missing steps for completing the operational concept is proposed, orderly defining the missing steps and possible solutions.

Introduction

In a global scenario of constantly increasing air traffic volumes [1], the aviation sector's main actors are redefining the entire air traffic management (ATM) concept, researching mitigating solutions to the current air-mobility issues, mainly related to crowded routes. A major bottleneck of the flight chain is aerodrome traffic, managed through the airport control tower, whose operators are responsible for the aerodrome traffic zone and on-ground, arrival, and departure operations. Aerodrome controllers comply with many aircraft manoeuvring with short time and space separations, relying on the direct out-of-the-window view to provide flight crews with essential clearances and flight instructions. Meanwhile, they retain important information from an increasing number of interfaces and coordinate with other controllers from the same or contiguous airspace sectors. Their work is thus characterised by a high – and still increasing – workload, while they need as much situational awareness as possible. Moreover, the requirements for improved safety of operations clash with the need for increased airport capacity and fluidity of traffic flow. Following the dictates of the fourth industrial revolution [2], research has focused on exploiting modern advanced human-machine interfaces – especially augmented/mixed reality (AR). AR exploit the superposition of an overlay of digital elements over the real world, providing the user with tailored information about the surrounding environment (thus "augmented" reality). Since AR usefulness requires matching the digital overlay and the physical world behind it, a critical aspect of AR techniques is the registration process, i.e. the continuous and real-time projection of the holograms in the correct position relative to the real world and the user's gaze.

By exploiting AR, operators should experience a decrease in the workload and an increase in situational awareness due to the possibility of directly retaining the information needed without gazing away from the outside view. This should result in an overall increase in the safety and fluidity of the traffic, also speeding up operations in low-visibility and hazardous conditions.

The first studies on the use of AR in airport control towers were conducted by Reisman in 2006 [3] and Tavanti [4] in 2007. They stated the need to include users in the design of AR tower applications while recognising that the technologies were still lacking in maturity. In 2015, following a considerable increase in AR maturity, Silva et al. [5] compared and integrated multiple surveillance sources (radar and ADS-B) to point the aircraft position on a video-based AR interface. In 2016, Gürlük [6] evaluated the beneficial effects of AR on situational awareness,

Aerospace Science and Engineering - III Aerospace PhD-Days   Materials Research Forum LLC
Materials Research Proceedings 42 (2024) 173-177   https://doi.org/10.21741/9781644903193-38

recognising some critical aspects of the ergonomics of the devices and the data organisation and visualisation. In 2018 [7], Gürlük et al. tested the implementation of AR in a virtual traffic scenario simulation involving tower controllers.

Starting in 2016, research campaigns have also been conducted at the University of Bologna. In 2019 and 2020, Bagassi et al. [8-10] presented the results of the SESAR-founded RETINA project, validating in a simulated scenario a fully operational concept which allowed the controllers to retrieve all the needed surveillance information in a head-up position, with an increase in task performance and situational awareness and reduction in workload. The concept tested two different projecting technologies – head-up displays and spatial displays – and identified see-through HMD displays as the best technology, with spatial displays still not presenting the needed maturity for dealing with multiple controllers. Simulations were selected for their cost-effectiveness, risk-safeness, and controllability, as already proven in many aerospace applications, including aircraft and helicopter flight simulators [11].

Following RETINA, the SESAR DTT project [12] evolved the developed concept, evaluating different solutions in simulated airport scenarios between 2021 and 2023. The proposed solutions included a label pinpointing the aircraft's position and presenting the needed surveillance information to the tower controller. The research assessed the adaptation of the interface to different environmental conditions and controller roles and validated different interaction modes with the surveillance infrastructure. Exhaustive validation campaigns were conducted using a human-in-the-loop methodology, employing several tower controllers in real time. The overall feedback confirmed a reduced workload and increased situational awareness. The DTT project increased the RETINA TRL, reaching level 4, thus preparing the developed concepts for operational implementation. This paper summarises the work done at the University of Bologna in transferring DTT solutions to an operational tower environment, hence defining a roadmap for the upcoming development and pointing out the technical challenges.

Challenges of AR application for real-time data in the control tower

In dealing with AR devices, the actual "augmentation" of the physical world implies an adequate matching of the holographic information with the user's view of the physical world. AR devices such as the Microsoft HoloLens 2™ head-mounted display (HMD), used at the University of Bologna's Virtual Reality and Simulation Lab, establish a visual link with the circumstancing environment through an array of cameras and depth sensors and place holograms accordingly to the scanned, detectable surrounding. While this is easier for a simulated scenario, where aircraft are visualised on screens locally placed in the room, real aircraft are usually detected through radar or transponder, which provide a global reference for the aircraft's position.

Similarly, much surveillance information, as particular indications of flight phases (push-back, taxiing…), is usually decided "a priori" somewhere alongside the simulation span rather than computed from available data. In real cases, the available information is limited, and the elaboration of surveillance data is mandatory to extract some precise information.

Also, the environmental conditions change between a simulated environment and actual airport rooms, with their out-of-the-window view and changing light and visibility conditions.

These considerations bring a new set of requirements. The user must retain its ability to move through the environment, so the AR device must track the user's gaze and position as well as the environment in real-time, repositioning the holograms simultaneously (real-time registration). Concurrently, the AR visualising device must be able to track aircraft based on global positioning data (usually related to an Earth-based reference system). These data should be converted to the AR device reference frame. Also, a source for the aircraft's surveillance data is needed, be it an antenna or a live link with third-party surveillance databases. These data come at uncertain rates, and missing data is possible. Sources for other relevant information, such as weather data and flight plans, are needed. Furthermore, unprecedented scenarios are possible, including

Aerospace Science and Engineering - III Aerospace PhD-Days   Materials Research Forum LLC
Materials Research Proceedings 42 (2024) 173-177   https://doi.org/10.21741/9781644903193-38

unexpected/hazardous traffic events or discrepancies in data flow content, aspect, etc. Finally, application development depends on the available traffic at the testing location.

Methodologies for AR implementation in the control tower. Due to the different implementation challenges, two methods are needed to design and validate the solutions developed for an operational tower environment. Technical checks and extended testing are required to ensure the conformal behaviour of the AR platform logic. A user-centred approach is needed instead for the interface design (view management, design of the graphics). This iterative method builds on previous validations, starting from the concepts assessed within the RETINA and DTT validation campaigns, and requires controllers' involvement in every design stage.

The current state of the physical-world implementation of AR in the control tower. Fadda et al. [13] developed a preliminary platform including initial solutions for many of the stated challenges. They designed a system for detecting the aircraft state vector and its positioning and rendering in the virtual world through the ADS-B surveillance technique, together with a primary organisation and representation of the information provided by the ADS-B in the digital tracking labels. The system is responsible for the correct superposition of the digital overlay onto the real-world scenario using a HoloLens2 see-through HMD. The central aspect of this solution - the capability of tracking aircraft in real-time and positioning the label with reasonable accuracy – has been validated with a technical check and the supervision of a tower controller. In addition, databases for airport weather and flight information are queried, and simple analysis of positional data for flight status detection (landing, take-off, taxiing or parked aircraft) is performed.

The developed application is thus capable of detecting aircraft movement and correctly collimating the holographic aircraft pointers and the surveillance information with the user's and aircraft's respective positions. The application can filter airport-relevant traffic among all the signals detected by the ADS-B antenna. It can also distinguish landing and departing aircraft, recognise their position in the airport and current flight status, and detect when they have parked.

An on-site validation of the concept developed so far has been recently performed at the Bologna Marconi airport (Fig. 1). The validation demonstrated the feasibility and reliability of the solution. Feedback from tower controllers was collected to guide future development.

Figure 1. User's view from the Bologna Marconi airport control tower.

A roadmap for future work in AR platform implementation

Starting from initial feedback collected during the demonstration at the Marconi airport, a roadmap is proposed to guide the future transfer to the physical world of the concepts developed in a simulated and fully controllable environment.

Aerospace Science and Engineering - III Aerospace PhD-Days   Materials Research Forum LLC
Materials Research Proceedings 42 (2024) 173-177   https://doi.org/10.21741/9781644903193-38

Preliminary feedback from tower controllers was on the device maturity (label visibility, display transparency), the interface design (missing adaptivity of label content, relevant information saliency, dimension of far aircraft pointer and visual clutter/label overlapping), and the data quality (system calibration, aircraft positional information delay).

The proposed future work collects feedback and identifies some known limitations. It should be organised as follows.

Improvement of the registration process for real-time data. The first task should be to improve the registration performance. This can be accomplished by working on the calibration procedure for aligning the local and global reference frames and on the accuracy and timeliness of the surveillance data. The calibration accuracy depends on determining the user's position inside the room, the room's position in the world, and the aircraft's position in the world. For the tracking of the user, marker and markerless methods should be compared. Room geodetic position can be determined through GPS sensor measures. Once the room registration accuracy has been perfected, room position and orientation can be passively improved by manually aligning reference holograms with known points. Finally, the live aircraft positioning quality depends on the accuracy of the ADS-B data received, and improvements are to be made to the conversion method between geodetic data and the AR device reference frame. Further progress comes from corrections for data transmission delay and predictive algorithms for data holes in the data feed.

Integration of surveillance and traffic management information. A significant aspect of the application development is integrating all the needed surveillance information, usually provided to controllers through the control tower's head-down interfaces. In addition to the challenging acquisition of this information, all the data should be managed to provide the controller with a usable and practical interface, choosing the correct data and extrapolating advanced information as ground flight phases, potentially implementing AI-based solutions. This step falls into a user-centred design, requiring an on-site validation with tower controllers. In contrast, the actual design of the interface could still take advantage of the rapid development allowed by simulated scenarios.

Integration of novel airport traffic control solutions. Tracking labels, attention guidance, and safety nets are novel concepts developed within the SESAR research framework. They should be integrated into the platform with concepts developed for DTT in a simulated scenario, such as ground phase recognition (push-back, start-up), air gestures, attention guidance and multimodal interaction [12]. This task will parallel the previous one since it requires reworking the available surveillance data.

Conclusion

The proposed roadmap highlights the main tasks identified for the completion of an operative platform to validate AR technologies inside the airport control tower. Further work could consider integrating Advanced Air Mobility traffic within the controller's workflow. This will require high levels of automation and proper management of the data provision to the controllers.

Finally, considering the recent advancements in the digital/remote tower concept, it is worth highlighting how the solutions proposed in this paper will still be valuable for a remote tower when considering multiple control roles (and thus points of view) and particular environmental conditions, such as low visibility.

Acknowledgements

Financed by the European Union - NextGenerationEU through the Italian Ministry of University and Research under PNRR - Mission 4 Component 1, Investment 4.1 (DM 118/2023) "PNRR Research".

References

[1] S. Shparberg, B. Lange, *Global Market Forecast 2022*, Airbus, Toulouse, 8th of July, 2022.

Aerospace Science and Engineering - III Aerospace PhD-Days   Materials Research Forum LLC
Materials Research Proceedings 42 (2024) 173-177   https://doi.org/10.21741/9781644903193-38

[2] D. Hanus, P. Revel, F. Marulo *et al.* "Aviation and the 4th Industrial Revolution: The Prominent Role of Networks of Excellence," *Aerotecnica Missili & Spazio*, vol. 96, pp. 86–89, 2017. https://doi.org/10.1007/BF03404740

[3] R. Reisman, D. Brown, "Design of Augmented Reality Tools for Air Traffic Control Towers," *6th AIAA Aviation Technology, Integration and Operations Conference, (ATIO)*, Wichita, Kansas, 25-27 September 2006. https://doi.org/10.2514/6.2006-7713

[4] M. Tavanti, "Augmented reality for tower: using scenarios for describing tower activities," *2007 IEEE/AIAA 26th Digital Avionics Systems Conference*, Dallas, TX, USA, 2007, pp. 5.A.4-1-5.A.4-12. https://doi.org/10.1109/DASC.2007.4391928.

[5] A. Silva, E. Ferreira, N. Laforge, S. Carvalho, "The use of surveillance data in augmented reality system to assist control tower operations," *2015 3rd Experiment International Conference (exp.at'15)*, Ponta Delgada, Portugal, 2015, pp. 178-182. https://doi.org/10.1109/EXPAT.2015.7463261.

[6] H. Gürlük, "Concept of an adaptive augmented vision based assistance system for air traffic control towers," *2016 IEEE/AIAA 35th Digital Avionics Systems Conference (DASC)*, Sacramento, CA, USA, 2016, pp. 1-10. https://doi.org/10.1109/DASC.2016.7777975.

[7] H. Gürlük, O. Gluchshenko, M. Finke, L. Christoffels and L. Tyburzy, "Assessment of Risks and Benefits of Context-Adaptive Augmented Reality for Aerodrome Control Towers," *2018 IEEE/AIAA 37th Digital Avionics Systems Conference (DASC)*, London, UK, 2018, pp. 1-10. https://doi.org/10.1109/DASC.2018.8569859.

[8] N. Masotti, S. Bagassi, F. De Crescenzio, "Augmented Reality for the Control Tower: The RETINA Concept," in *De Paolis, L., Mongelli, A. (eds) Augmented Reality, Virtual Reality, and Computer Graphics, AVR 2016*. Lecture Notes in Computer Science, vol. 9768. Springer, Cham, 2016. https://doi.org/10.1007/978-3-319-40621-3_32

[9] N. Masotti, F. De Crescenzio, S. Bagassi, "Augmented Reality in the Control Tower: A Rendering Pipeline for Multiple Head-Tracked Head-up Displays," in *De Paolis, L., Mongelli, A. (eds) Augmented Reality, Virtual Reality, and Computer Graphics, AVR 2016*. Lecture Notes in Computer Science, vol. 9768. Springer, Cham, 2016. https://doi.org/10.1007/978-3-319-40621-3_23

[10] S. Bagassi, F. De Crescenzio, S. Piastra, C.A. Persiani, M. Ellejmi, A.R. Groskreutz, J. Higuera, "Human-in-the-loop evaluation of an augmented reality-based interface for the airport control tower," in *Computers in Industry*, vol. 123, 103291, 2020. https://doi.org/10.1016/j.compind.2020.103291

[11] M. Daniele, G. Quaranta, P. Masarati *et al.* "Pilot in the Loop Simulation of Helicopter-Ship Operations Using Virtual Reality", in *Aerotecnica Missili e Spazio*, vol. 99, pp. 53–62, 2020. https://doi.org/10.1007/s42496-020-00037-3

[12] R. Santarelli, S. Bagassi, M. Corsi, J. Teutsch, R. Garcia Lasheras, M.A. Amaro Carmona, A.R. Groskreutz, "Towards a digital control tower: the use of augmented reality tools to innovate interaction modes", *12th SESAR Innovation Days*, Budapest, Hungary, 5-8 December 2022. https://www.sesarju.eu/sesarinnovationdays

[13] T. Fadda, S. Bagassi, M. Corsi, "ADS-B driven implementation of an augmented reality airport control tower platform", in *Materials Research Proceedings*, vol. 37, pp. 767-770, 2023. https://doi.org/10.21741/9781644902813-164

Aerospace Science and Engineering - IV Aerospace PhD-Days   Materials Research Forum LLC
Materials Research Proceedings 42 (2024) 178-182   https://doi.org/10.21741/9781644903193-39

Deep learning-based spacecraft pose estimation for the pre-capture phase scenario

Roman Prokazov[1,a*]

[1]Department of Industrial Engineering, Alma Mater Studiorum Università di Bologna Via Fontanelle 40, 47121, Forlì, Italy

[a]roman.prokazov2@unibo.it

Keywords: Deep Learning, Computer Vision, Spacecraft Pose Estimation, Synthetic Data

Abstract. The increasing population of space debris in Low-Earth Orbit (LEO) poses a significant threat to operational satellites and future space endeavors. To address this challenge, leading aerospace companies worldwide are developing on-orbit servicing and debris removal satellites. These servicer satellites will be capable of complex orbital operations, such as capturing tumbling defunct spacecraft. A fundamental requirement for the success of such missions is the development of accurate spacecraft pose estimation, which provides the servicer's guidance and control system with precise information about the target spacecraft's attitude. This paper addresses the study of such a pipeline using deep learning and classical computer vision algorithms.

Introduction

The accurate estimation of the relative position and attitude (i.e., pose) of a spaceborne object using minimal hardware is an enabling technology for current and future on-orbit servicing and debris removal missions [1]. These missions play a crucial role in mitigating the growing problem of space debris in Low-Earth Orbit (LEO) [2]. Among them are RemoveDEBRIS by the Surrey Space Centre [3], Clearspace-1 by the Swiss start-up 'Clearspace SA' [4], CRD2 by JAXA [5], and EROSS by Thales Alenia Space [6]. Efficient mission planning for such operations requires trajectory optimization techniques, especially when considering refueling or resupply scenarios involving in-situ resource utilization (ISRU) on celestial bodies like the Moon [7]. These complex maneuvers necessitate not only meticulously planned trajectories but also real-time awareness of the target spacecraft's pose throughout the mission. By providing crucial information about the target spacecraft's pose relative to the servicing spacecraft, pose estimation systems serve as the critical "eyes" for the mission. This information is continuously fed into the guidance system. The guidance system utilizes the pose data alongside the pre-programmed mission plan and real-time sensor data to constantly adjust the spacecraft's thrusters, ensuring it stays on the optimized trajectory and executes maneuvers with high precision [8]. Depending on the scenario, the target spacecraft can be either cooperative, utilizing radio links or fiducials, or non-cooperative with known or unknown geometry. Recently, non-cooperative target pose estimation has gained interest from the aerospace society due to the accumulation of inactive satellites and debris in low Earth orbit. However, the current go-to solution for performing pose estimation of target spacecraft involves cumbersome and expensive LIDARs, which will likely hinder the widespread adoption of space debris removal technology. One of the most promising alternatives to LIDARs to date is the use of monocular cameras combined with computer vision algorithms, providing a cheap and lightweight solution. To address this, ESA organized the international Satellite Pose Estimation competition (SPEC2021), the main objective of which was to find the most efficient way to estimate the pose of known uncooperative spacecraft using monocular cameras [9]. SPEC2021 exploits the next-generation spacecraft pose estimation dataset (SPEED+), consisting of labeled images from computer graphics and real-life sources. To solve this problem, each participant implemented deep learning (DL) techniques as opposed to classical image processing methods. This choice stems from the fact that the latter are too computationally expensive for on-board processors and are not fully robust to the complex space environment, such as harsh lighting conditions. Neural networks, on the other hand, are able to learn complex features that are sometimes impossible to hardcode by humans. However, when trained on synthetic data, DL models typically fail when tested on real images, leading to the so-called domain gap problem [9].

This paper proposes a DL-based pipeline for the pose estimation of uncooperative target spacecraft with known geometry in the pre-capture phase prior to docking. Our method is based on state-of-the-art pose estimation network (YOLOv8-pose) [10] and Perspective-n-Point (PnP) solvers [11]. PnP solvers are well-established algorithms that efficiently calculate the 3D pose of an object in a scene given its corresponding 2D keypoints and a 3D model of the object. The scenario considered in this paper is a synthetic video stream of the client uncooperative spacecraft approaching the servicer until reaching a minimum relative distance of 20 cm, which corresponds to the length of the servicer gripper. To generate a realistic training dataset for the deep learning network we utilized Blender, a popular open-source 3D creation suite [12], and CAD model of the Envisat spacecraft [13]. The Envisat model in Blender was then animated to simulate the approach of a client spacecraft towards a servicer spacecraft, mimicking the pre-capture phase prior to docking.

Methods

Data preparation. The generation of synthetic images is realized through the open-source 3D graphical tool Blender. Firstly, this choice is made because of its integrated Cycles rendering engine, which is physically based and uses ray tracing. Secondly, Blender offers a built-in Python API, enabling control over the software via Python scripting. Another useful tool utilized is the third-party Starfish Python library, which facilitates the automatic generation of thousands of images along with pose information for the object in each image. Users can define the object's pose or have it randomly distributed across frames. Moreover, the Starfish library allows for the variation of parameters such as background and lighting orientation in each generated image. After the image dataset is generated with corresponding annotations regarding the position and orientation of the spacecraft, it is necessary to preprocess the dataset and add additional information to the annotation file to enable its use in training neural networks. Specifically, knowing the 3D coordinates of the spacecraft and its pose information, it is possible to retrieve the pixel coordinates of each keypoint using perspective projection equation:

$$\begin{bmatrix} u_i w_i \\ v_i w_i \\ w_i \end{bmatrix} = \begin{bmatrix} f_x & 0 & c_x \\ 0 & f_y & c_y \\ 0 & 0 & 1 \end{bmatrix} [R(\mathbf{q_{BC}}) \ \mathbf{t_{BC}}] \begin{bmatrix} x_B \\ y_B \\ z_B \end{bmatrix} \tag{1}$$

where:

$-x_B, y_B, z_B$ are the 3 d coordinates of the spacecraft keypoint

$-[R(\mathbf{q_{BC}})\mathbf{t_{BC}}]$ is roto-translation vector

$-f_x, f_y, c_x, c_y$ are the parameters of the intrinsic matrix
$-u_i w_i, v_i w_i$ are the pixel coordinates of the spacecraft key-point

Pose estimation pipeline. The pipeline can be summarized in 2 following steps according to the figure 1:

1. Deployment of Keypoint Regression Network (KRN). In this paper we decided to leverage ultralytics API which uses YOLOv8-pose neural network architecture. It gives the users to choose

Aerospace Science and Engineering - III Aerospace PhD-Days
Materials Research Proceedings 42 (2024) 178-182

Materials Research Forum LLC
https://doi.org/10.21741/9781644903193-39

between different model size and can be easily transformed to openvvino format for the faster inference on the low-power compute devices. The KRN yields a $1 \times 2N$ vector encoding the 2D positions of N keypoints.

2. Retrieved 2D keypoints together with the available 3d wireframe model of the spacecraft can be fed into off the-shelf PnP solver in order to calculate the position and orientation of the spacecraft.

The selected ground-truth keypoints were strategically chosen to ensure their presence within the field of view (FOV) of the monocular camera. Moreover, a minimum of four points is required to execute the PnP (Perspective-n-Point) algorithm effectively.

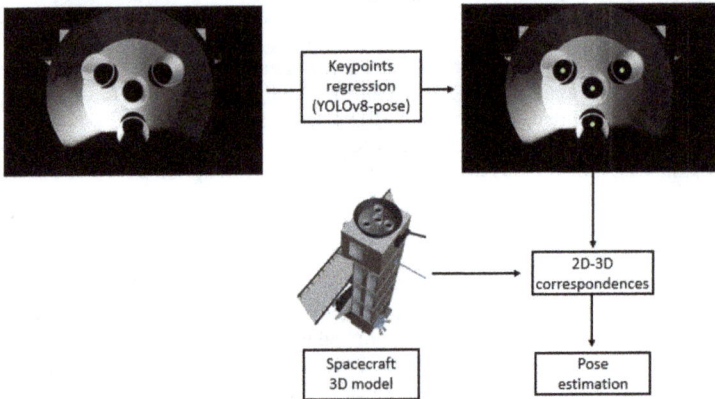

Figure 1: 2-stage pose estimation pipeline.

On the figure 2 there is visualization of some images from the validation test set. It is evident that the KRN based on YOLOv8-pose architecture is able to output the right posses with the very high probability.

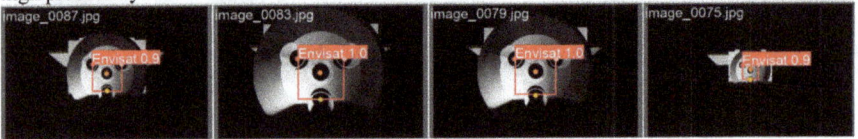

Figure 2: Samples from validation batch during the training of KRN.

The retrieved poses of the spacecraft are compared to the ground truth data according to the metrics from the SPEC2021 competition:

$$E = \frac{1}{N}\sum_{i=1}^{N}\left(\frac{|t_{BC_i} - \hat{t}_{BC_i}|_2}{|t_{BC_i}|_2} + 2\arccos(|\langle q_i, \hat{q}_i\rangle|)\right) \tag{2}$$

Where t_{BC_i}, q_i and $\hat{t}_{BC_i}$, $\hat{q}_i$ are the ground truth and estimated position vectors and attitude quaternions respectively.

Results

A dataset comprising 5000 images from Envisat, each with a resolution of 1920×1200, was created. This dataset was then partitioned into three subsets: a training set containing 3500 images,

Aerospace Science and Engineering - III Aerospace PhD-Days　　　　　Materials Research Forum LLC
Materials Research Proceedings 42 (2024) 178-182　　　　　https://doi.org/10.21741/9781644903193-39

a validation set containing 1000 images, and a test set containing 500 images. Considering the limitations of low-power on-board spacecraft processors, we opted for the smallest YOLOv8n-pose network model available from the ultralytics API (See table 1).

Table 1: Characteristics of YOLOv8-pose models with different sizes

Model	size (pixels)	mAP^{pose} 50-95	mAP^{pose} 50	Speed CPU ONNX (ms)	Speed A100 TensorRT (ms)	params (M)	FLOPs (B)
YOLOv8n-pose	640	50.4	80.1	131.8	1.18	3.3	9.2
YOLOv8s-pose	640	60.0	86.2	233.2	1.42	11.6	30.2
YOLOv8m-pose	640	65.0	88.8	456.3	2.00	26.4	81.0

Training was performed on Nvidia GeForce RTX 3090 for 100 epochs with the minimum train and validation pose loses of 0.055 and 0.033 respectively (figure 3). Batch size was set of 32 images.

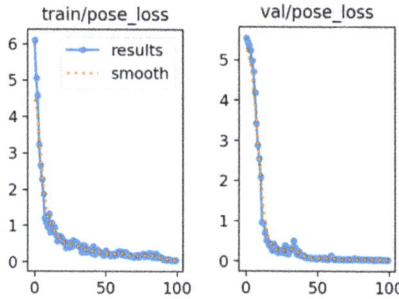

Figure 3: Train and validation pose losses.

The predicted keypoints for each image frame were then fed into the PnP solver to retrieve the poses for the spacecraft across the test set of 500 images dedicated to final validation. Preliminary results are presented in Table 2 and will be improved upon in future work.

Table 2: Evaluation metrics according to the SPEC2021

Mean translation error e_t (m)	0.01907
Mean rotation error e_q (deg)	14.89
Mean SPEED score E	0.29

References

[1] T. H. Park, S. Sharma, and S. D'Amico, "Towards robust learning-based pose estimation of noncooperative spacecraft," arXiv preprint arXiv:1909.00392, 2019

[2] Pietro, D.M. A Ptolemaic Approach Improving the Conjunction Analysis Pipeline for Leo. Aerotec. Missili Spaz. 102, 309–321 (2023). https://doi.org/10.1007/s42496-023-00164-7

[3] J. L. Forshaw, G. S. Aglietti, N. Navarathinam, H. Kadhem, T. Salmon, A. Pisseloup, E. Joffre, T. Chabot, I. Retat, R. Axthelm, et al., "Removedebris: An in-orbit active debris removal demonstration mission," Acta Astronautica, vol. 127, pp. 448–463, 2016. https://doi.org/10.1016/j.actaastro.2016.06.018

[4] R. Biesbroek, S. Aziz, A. Wolahan, S.-f. Cipolla, M. Richard-Noca, and L. Piguet, "The clearspace-1 mission: Esa and clearspace team up to remove debris," in Proc. 8th Eur. Conf. Sp. Debris, pp. 1–3, 2021.

[5] T. Yamamoto, J. Matsumoto, H. Okamoto, R. Yoshida, C. Hoshino, and K. Yamanaka, "Pave the way for active debris removal realization: Jaxa commercial removal of debris demonstration (crd2)," in 8th European Conference on Space Debris, p. 200, 2021.

[6] V. Dubanchet, S. Andiappane, P. Negro, D. Casu, A. Giovannini, G. Durand, and J. D'Amico, "Validation and demonstration of eross project: The european robotic orbital support services," in Proceedings of the International Astronautical Congress, IAC, 2020.

[7] Siena, A. Orbit/Attitude Control for Rendezvous and Docking at the Herschel Space Observatory. Aerotec. Missili Spaz. 103, 39–49 (2024). https://doi.org/10.1007/s42496-023-00188-z

[8] Atmaca, D., Pontani, M. Near-Optimal Feedback Guidance for Low-Thrust Earth Orbit Transfers. Aerotec. Missili Spaz. (2024). https://doi.org/10.1007/s42496-023-00193-2

[9] T. H. Park, M. Märtens, M. Jawaid, Z. Wang, B. Chen, T.-J. Chin, D. Izzo, and S. D'Amico, "Satellite pose estimation competition 2021: Results and analyses," Acta Astronautica, vol. 204, pp. 640–665, 2023. https://doi.org/10.1016/j.actaastro.2023.01.002

[10] Information on https://docs.ultralytics.com/tasks/pose/

[11] Lu, Xiao. (2018). A Review of Solutions for Perspective-n-Point Problem in Camera Pose Estimation. Journal of Physics: Conference Series. 1087. 052009. https://doi.org/10.1088/1742-6596/1087/5/052009

[12] Information on https://www.blender.org/

[13] CAD model: https://sketchfab.com/3d-models/envisat-65b0ec49681a44f68dfc8bd4efe9583

Aerospace Science and Engineering - IV Aerospace PhD-Days
Materials Research Proceedings 42 (2024) 183-187

Materials Research Forum LLC
https://doi.org/10.21741/9781644903193-40

Space-based observation modeling method considering resident space object state uncertainty

Yifan Cai[1,a*], Camilla Colombo[2]

[1]Ph.D Student, Politecnico di Milano, Via La Masa 52, Milano, Italy

[2]Associate Professor, Politecnico di Milano, Via La Masa 52, Milano, Italy

[a]yifan.cai@polimi.it

Keywords: Space-based Observation, Resident Space Objects, Uncertainty, Monte Carlo

Abstract. The proliferation of Resident Space Objects (RSOs) is accelerating at an unprecedented pace, leading to a notable escalation in RSO density and heightened risks to space operations. Consequently, there is an urgent demand for more effective methodologies to estimate the orbital state of RSOs, ensuring the sustainability of space activities. Compared with ground-based observation, space-based observation systems offer significant advantages in accuracy and timeliness owing to their capability for continuous and comprehensive monitoring. However, space-based observation sensors have limited field of view, resulting in their ability to accommodate only a few number of RSOs. Therefore, accurately assessing the detectability of RSOs relative to the sensor and using this information to determine which RSOs to observe is crucial. This can prevent wastage of limited sensor observation resources, thereby acquiring more effective observational data to attain precise RSO states. Due to the inevitable uncertainty in the characterization of RSO states, solely relying on certainty geometric relationships to calculate the detectability is not sufficiently accurate. This paper proposes a space-based observation modeling method that incorporates the uncertainty of the RSO state. By integrating the Monte Carlo (MC) random sampling idea, this method is developed based on probabilistic relationships. This enables the derivation of a numerically continuous distribution of the detectable probability of RSOs, rather than only focusing on whether they are detectable. The intriguing aspect of this research lies in embedding the uncertainty of RSO states into all stages of estimation, thereby enhancing the accuracy of RSO state estimation under limited sensor observation resources.

Introduction

In a space object tracking system, the intrinsic connection between a space object and a moment is the state change of the space object over moments. The space object tracking problem can be modelled as a correspondence between the observation space and the state space of the space object. In the state space, the state of a space object is shifted from the previous moment $k-1$ to the next moment k. However, in the observation space, since only part of the space object state can be observed, a Bayesian recursive method[1] is used to estimate the space object state to obtain a set of space object state information over the moments[2]. Especially with the rapid increase of mega constellations, more RSO state estimates require more efficient us of limited sensor resources[3]. In this process, precisely determining whether a space object can be successfully detected by sensor is essential and challenging.

This research presents a space-based observation modeling approach considering the uncertainty of the space object state. Based on the traditional model relied on geometric relationships, a model reflecting probabilistic relationships is developed through the integration of the Monte Carlo random sampling idea[4]. Compared to traditional space-based observation models, this method embeds the uncertainty of RSO states into various stages of estimation, while also generating a continuous numerical distribution representing the detectability probability of space objects. This enhances the robustness of space object state estimation[5].

Space-based observation model based on geometric relationships
Observation sensors typically employ two operational strategies: staring and tracking modes[6]. Both require assessing the detectability of space objects. The process of modelling space-based observations based on geometric relationships is shown in Figure 1.

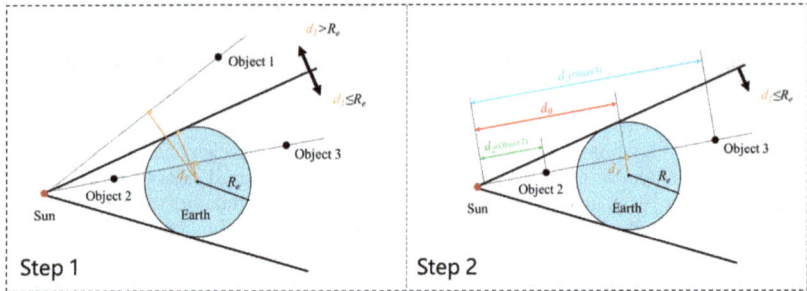

Figure 1 Process for calculating whether a space object is covered by the Earth's shadow.

Whether a space object can be observed by the sensor within the time window is determined by the visibility $p_{v,k}$ of the space object relative to the sensor, which is calculated as

$$p_{v,k} = (1-p_{s,k})(1-p_{b,k}) \qquad (1)$$

where $p_{s,k}$ indicates whether the space object is covered by the Earth's shadow, obtained from the geometrical relationship between the Sun, the space object and the Earth, and $p_{s,k}$ is 0 when the space object is not in the shadow formed by the Sun on the Earth, and 1 otherwise; and $p_{b,k}$ indicates whether the space object is occluded by the Earth, obtained from the geometrical relationship between the sensor, the space object and the Earth, and $p_{b,k}$ is 0 if the space object is not occluded by the Earth in the field of view of the sensor, and 1 otherwise.

The calculation of $p_{s,k}$ and $p_{b,k}$ is defined as the "0-1 model". The calculation of $p_{s,k}$ consists of two steps. The first step defines the length of the vertical line from the centre of the Earth to the line between the Sun and the space object as d_1, the radius of the Earth as R_e, then $p_{s,k}$ is given by

$$p_{s,k} = \begin{cases} 0, & d_1 > R_e \\ 0 \text{ or } 1, & d_1 \le R_e \end{cases} \qquad (2)$$

When $d_1 > R_e$, the space object is not in the shadow formed by the Sun on the Earth, i.e. $p_{s,k} = 0$. However, when $d_1 \le R_e$, whether the space object is in the shadow zone will depend on its position relative to the Sun and the Earth. Define the distance from the center of the Sun to line d_1 as d_0, and the distance from the center of the Sun to the space object as d_2. Comparing the lengths of d_2 and d_0 to determine whether the space object is shadowed by the Earth, i.e.

$$p_{s,k} = \begin{cases} 0, & d_2 < d_0 \\ 1, & d_2 > d_0 \end{cases} \qquad (3)$$

Aerospace Science and Engineering - III Aerospace PhD-Days   Materials Research Forum LLC
Materials Research Proceedings 42 (2024) 183-187   https://doi.org/10.21741/9781644903193-40

The calculation of $p_{b,k}$ consists of two steps. The first step defines the length of the vertical line from the centre of the Earth to the line between the sensor and the space object as d_1, the radius of the Earth as R_e, then $p_{b,k}$ is expressed as follows.

$$p_{b,k} = \begin{cases} 0, & d_1 > R_e \\ 0 \text{ or } 1, & d_1 \le R_e \end{cases} \quad (4)$$

When $d_1 > R_e$, the space object is not in a region that may be occluded by the Earth, i.e. $p_{s,k} = 0$. However, when $d_1 \le R_e$, whether the space object is in an Earth-obscured region will depend on its position relative to the sensor and the Earth. Define the distance from the sensor to line d_1 as d_0, and the distance from the sensor to the space object as d_2. Comparing the lengths of d_2 and d_0 to determine whether the space object is occluded by the Earth, i.e.

$$p_{b,k} = \begin{cases} 0, & d_2 < d_0 \\ 1, & d_2 > d_0 \end{cases} \quad (5)$$

Space-based observation model based on probabilistic relationships

Since the large-scale space object tracking problem uses a limited number of space-based optical sensors to track large-scale space' objects, it results in a large number of space objects frequently moving in and out of the sensors' field of view. Therefore, correctly calculating the detectability $p_{D,k}$ of a space object to measure the presence or absence of a space object at a given moment k is crucial for accurately estimating the number of space objects. The $p_{D,k}$ is time-varying due to changes in lighting conditions and the relative distance between the sensor and the space object. Therefore, more accurate $p_{D,k}$ prediction model is needed.

The detectability of a space object is determined only by its visibility relative to the sensor. In practical missions, the apparent magnitude of a space object relative to the sensor has a significant impact on its observation quality. The $p_{D,k}$ consists of three components, i.e., the probability of successful detection of a space object as p_D, the apparent magnitude of the space object relative to the sensor as $p_{am,k}$, and the visibility of the space object relative to the sensor as $p_{v,k}$, which is calculated as

$$p_{D,k} = p_D p_{am,k} p_{v,k} \quad (6)$$

where p_D is a constant value that measures the performance of the sensor itself over $[0,1]$; $p_{am,k}$ is calculated from the relative positions of the space object, the sensor and the sun. $p_{v,k} = (1 - p_{s,k})(1 - p_{b,k})$. $p_{s,k}$ and $p_{b,k}$ are determined by the "0-1 model".

Based on the uncertainty of the state of the space object and the probability theory, the "0-1 model" of $p_{s,k}$ and $p_{b,k}$ is refined into the "0~1 model" of $p_{s,k}^{prob}$ and $p_{b,k}^{prob}$ in order to obtain a more accurate $p_{D,k}^{prob}$ with respect to the $p_{D,k}$. Where $p_{s,k}^{prob}$ and $p_{b,k}^{prob}$ are the probabilities of the space object being in the shadow of the Earth and being obscured by the Earth, respectively, and the range of values are $[0,1]$.

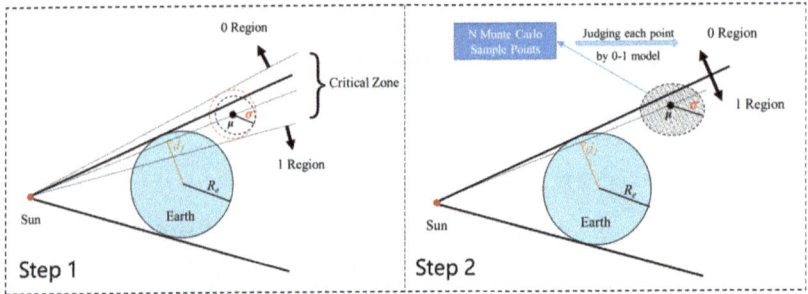

Figure 2 Process for calculating the probability that a space object is covered by the Earth's shadow.

The calculation of $p_{s,k}^{prob}$ requires two steps as shown in Figure 2. The first step defines the maximum two-dimensional standard deviation of the three-dimensional covariance of the space object in the plane of the Sun, the Earth, and the space object as σ. Since the distance between the Earth and the Sun is much larger than the distance between the Earth and the space object, it is approximated whether the space object is in the critical zone by comparing the magnitude relationship between $|d_1 - R_e|$ and $|\sigma|$. Then, the $p_{s,k}^{prob}$ is expressed as

$$p_{s,k}^{prob} = \begin{cases} 0, & |d_1 - R_e| > |\sigma| \text{ and } d_1 > R_e \\ p_{s,k}^{CR}, & |d_1 - R_e| \le |\sigma| \\ 1, & |d_1 - R_e| > |\sigma| \text{ and } d_1 < R_e \end{cases} \qquad (7)$$

where $p_{s,k}^{CR}$ is the probability that the space object in the critical zone is in the shadow of the Earth. It can be found that if $|d_1 - R_e| \le \sigma$, then the space object may be in the shadowed or non-shadowed area. Therefore, it is necessary to further calculate $p_{s,k}^{CR}$ for the space objects in critical zone.

Considering the state uncertainty of the space object and the computational complexity, the probability that the space object is in the shadow of the Earth is calculated by combining the Monte Carlo sampling method. For a space object j, N sample points satisfying Gaussian distribution $N(\mu, \sigma^2)$ are randomly generated in its state distribution space, each sample point represents a space object, and each space object may be located in the shadow area or non-shadow area. From Eq. (2) and Eq. (3), there are N_o sample points located in the shaded area and N_u sample points located in the non-shaded area. Then $p_{s,k}^{CR}$ is calculated by the ratio of N_o and N as

$$p_{s,k}^{CR} = \frac{N_o}{N} \qquad (8)$$

Eq. (6) is transformed to

$$p_{D,k}^{prob} = p_D p_{am,k} p_{v,k}^{prob} \qquad (9)$$

where the probability that the space object is visible relative to the sensor is calculated as $p_{v,k}^{prob} = (1 - p_{s,k}^{prob})(1 - p_{b,k}^{prob})$.

Up to this point, Eq. (7)-Eq. (9) obtain the space object observation model of the space-based tracking system based on the probabilistic relationship, and obtain the space object detectability probability that is numerically continuously distributed, i.e., realize the prediction of space object detectability probability, and improve the accuracy of the state estimation of the space object compared to the existing method of judging the detectability of the space object that is discretely distributed.

Summary

The space-based observation model modelling method considering space object state uncertainty proposed in this paper establishes a space-based observation model based on probabilistic relationships by considering space object state uncertainty based on a space-based observation model based on geometric relationships. The space-based observation model established by this method can obtain numerically continuously distributed space object detectability probabilities, which improves the accuracy and robustness of space object state estimation compared with existing discrete-distributed space object detectability judgement methods.

References

[1] H.W. Sorenson, D.L. Alspach, Recursive Bayesian estimation using Gaussian sums, Automatica. 1971, 7(4): 465-479. https://doi.org/10.1016/0005-1098(71)90097-5

[2] H. Cai, S. Gehly, Y. Yang, et al, Multisensor tasking using analytical Rényi divergence in labeled multi-Bernoulli filtering, Journal of Guidance, Control, and Dynamics. 2019, 42(9): 2078-2085. https://doi.org/10.2514/1.G004232

[3] J. Zhang, Y. Cai, C. Xue, et al, LEO mega constellations: review of development, impact, surveillance, and governance, Space: Science & Technology. 2022. https://doi.org/10.34133/2022/9865174

[4] F. James, Monte Carlo theory and practice, Reports on progress in Physics. 1980, 43(9): 1145. https://doi.org/10.1088/0034-4885/43/9/002

[5] X. Qian, L. Li, J. Zhang, et al, Generic Homotopic Smoothing for Low-Thrust Optimal Control Problems With Power Constraints, IEEE Transactions on Control Systems Technology. 2024.

[6] J. G. López-Cepero, AstroDART: Astronomical data analysis and recovery from tracklets, Aerotecnica Missili & Spazio. 2023, 102(4): 355-365. https://doi.org/10.1007/s42496-023-00174-5

Aerospace Science and Engineering - IV Aerospace PhD-Days
Materials Research Proceedings 42 (2024) 188-192

Materials Research Forum LLC
https://doi.org/10.21741/9781644903193-41

Numerical characterisation of additively manufactured components

Giuseppe Mantegna[1,a] *, Davide Tumino[1,b], Calogero Orlando[2,c] and Andrea Alaimo[1,d]

[1]Kore University of Enna - Department of Engineering and Architecture, 94100, Enna (EN), Italy

[a]giuseppe.mantegna@unikore.it, [b]davide.tumino@unikore.it, [c]calogero.orlando@unikore.it, [d]andrea.alaimo@unikore.it

Keywords: Additive Manufacturing, FEM, Lattice

Abstract. Additive manufacturing (AM) technologies are gaining widespread adoption across multiple engineering sectors due to their ability to create customised components with tailored mechanical properties and reduced material waste. This work presents different innovative Finite Element Method (FEM) models to characterise the behaviour of AM structures. A numerical model is presented to characterise AM porosity defects using the Representative Volume Element (RVE) concept; additionally, different unit cell approaches to study the behaviour of lattice structures are presented. The procedures and methodologies presented offer a range of tools, each with different trade-offs between accuracy and computational cost.

Introduction

Additive Manufacturing (AM), commonly referred to as 3D printing, is a process that involves building objects by incrementally adding material layer-by-layer based on a Computer-Aided Design (CAD) model. One of the major advantages of AM is its ability to produce finished components using a single tool, and the capability to create complex geometric shapes while minimising material waste. Moreover, studies have highlighted the cost savings associated with AM, particularly for small-volume production runs, as well as a shorter design-to-development cycle time [1]. AM technologies have facilitated the integration of lattice structures into complex structural components, enabling weight optimisation while simultaneously maximising mechanical performance tailored to specific functional requirements. Indeed, the fabrication of such complex structures using traditional manufacturing techniques would be highly complex or infeasible. Despite the numerous advantages offered by AM technologies, it must be acknowledged that AM components can exhibit defects and imperfections that may adversely affect their mechanical properties. These defects can arise due to various factors, such as process parameters and/or material characteristics. Consequently, it is essential to be able to forecast the behaviour of AM components. Numerical models not only provide valuable insights into the structure-property relationships but also enable the optimisation of design parameters and manufacturing processes.

Finite Element Model Porosity Characterisation

Internal defects, such as porosity, are common in additively manufactured components, and they can lead to inferior mechanical properties, potentially compromising the structural integrity and safety of the entire system. To address this challenge, a numerical model based on the concept of the Representative Volume Element (RVE) is presented to predict the macroscopic behaviour of the structure. By capturing the microscale defects within the RVE, is possible to predict the macroscopic effects through a homogenization procedure, which significantly reduces the computational costs associated with the macro-scale analysis [2].

The study of Cai et al. [3] is considered as reference for the pore size distribution inside SLM AlSi10Mg components. The developed RVE model uses hollow spheres to simulate pores inside

Aerospace Science and Engineering - III Aerospace PhD-Days   Materials Research Forum LLC
Materials Research Proceedings 42 (2024) 188-192   https://doi.org/10.21741/9781644903193-41

the component. Isolated spheres represent metallurgical pores, while the shapes generated from their overlap simulate keyhole pores generated by the coalescence of multiple pores during the SLM manufacturing process. An in-house APDL (Ansys Parametric Design Language) code is developed to generate the RVE with pores, Figure 1.

a)

b)

c)

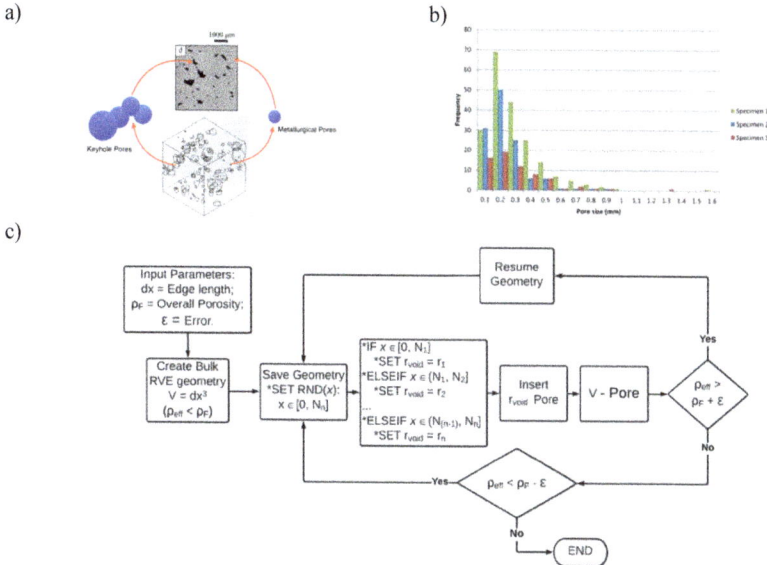

Figure 1 – RVE porosity scheme. a) Pores idealisation, b) Pores size distribution [3], c) APDL flow chart diagram.

Compression and shear numerical tests are performed on the RVE to characterise its mechanical behaviour. To ensure accurate results, double periodic boundary conditions are imposed by coupling the displacement of each node on two opposing faces that share the same relative position with two pilot nodes, Np. The periodic conditions along the x-direction are:

$$u_{i(0,y_i,z_i)} - u_{i'(dx,y_i,z_i)} = \Delta u_{Np}$$
$$v_{i(0,y_i,z_i)} - v_{i'(dx,y_i,z_i)} = \Delta v_{Np} \qquad (1)$$
$$w_{i(0,y_i,z_i)} - w_{i'(dx,y_i,z_i)} = \Delta w_{Np}$$

As expected, as porosity increases, key mechanical properties decrease, including the elastic modulus, the shear modulus, yield strength and Poisson's coefficient, Figure 2.

a)

b)

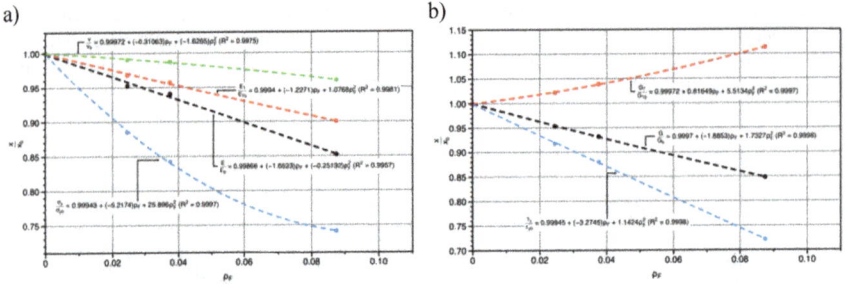

Figure 2 – Normalised properties respect with defect-free material [2]. a) Compression Test, b) Shear test.

Waved Body-Centred Cubic Cell

Lattice structures are formed by the ordered repetition of unit cells in three-dimensional space to create the final component. Among the different cell topologies, strut-based cells are often used to their manufacturing and modelling simplicity. The standard BCC cell has been extensively studied in the literature [4], [5]. A new variant of the BCC cell has been developed by the authors [6], wherein the straight struts of the traditional BCC cell are replaced by sinusoidal struts, making to the new cell orthotropic:

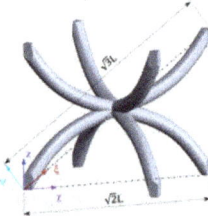

$$\psi = a \sin\left(\frac{2\pi\xi}{\sqrt{3}L}\right) \tag{2}$$

Figure 3 – Waved Body-Centred Cell.

Compression and shear tests have been conducted to characterise the behaviour of the new unit cell. To achieve accurate results, the double periodic boundary conditions employed for the RVE, Eq. (1), have been suitably modified. Additionally, a novel procedure has been developed to reduce the computational costs associated with the analysis. This approach utilizes simplified periodic conditions through the implementation of remote points to constrain the lateral faces, as shown in Figure 4.

a)

b)

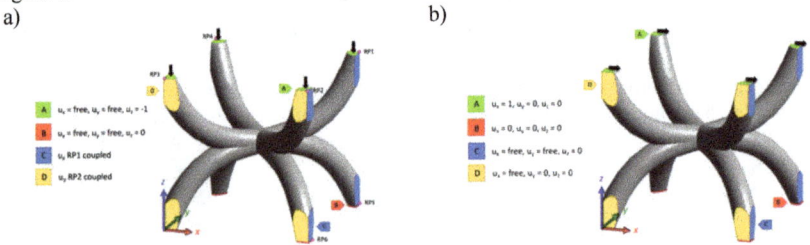

Figure 4 – Simplified periodic boundary conditions [6]. a) Compression test, b) Shear test.

Aerospace Science and Engineering - III Aerospace PhD-Days   Materials Research Forum LLC
Materials Research Proceedings 42 (2024) 188-192   https://doi.org/10.21741/9781644903193-41

Table 1 summarises the elastic modulus E_z results obtained with the double periodic boundary conditions (2xp) and the simplified periodic conditions (s-p). A good approximation can be outlined with a maximum error of only 2%. Table 2 summarises the mechanical properties of the waved unit cell.

Table 1 – WBCC double periodicity-simplified periodicity E_z modulus comparison.

ϕ	E_z 2xp	E_z s-p	e%
[mm]	[MPa]	[MPa]	
1.5	22.12	21.79	-1.5%
1.75	46.5	45.56	-2.0%
2	85.84	83.58	-2.6%
2.25	141.62	138.03	-2.5%
2.75	297.46	290.39	-2.4%

Table 2 – WBCC equivalent elastic properties.

ϕ	$E_x = E_y$	E_z	v_{12}	$v_{23} = v_{13}$	G_{12}	$G_{23} = G_{13}$
[mm]	[MPa]	[MPa]	[ ]	[ ]	[MPa]	[MPa]
1.5	7.35	22.12	0.721	0.175	38.74	7.33
1.75	14.99	46.5	0.723	0.155	54.54	12.83
2	27.92	85.84	0.719	0.138	73.37	20.51
2.25	48.12	141.62	0.705	0.125	103.97	30.62
2.75	120.62	297.46	0.653	0.1	158.11	58.84

Conclusions

Novel FEM models have been developed to characterise various aspects of additively manufactured structures. The proposed models offer designers a versatile toolset, providing the flexibility to select the optimal balance between accuracy and computational costs, thereby tailoring the approach to specific design requirements.

References

[1] V. Mohanavel, K. S. A. Ali, K. Ranganathan, J. A. Jeffrey, M. M. Ravikumar, and S. Rajkumar, "The roles and applications of additive manufacturing in the aerospace and automobile sector," *Mater Today Proc*, vol. 47, pp. 405–409, 2021. https://doi.org/10.1016/j.matpr.2021.04.596

[2] G. Mantegna *et al.*, "Representative volume element homogenisation approach to characterise additively manufactured porous metals," *Mechanics of Advanced Materials and Structures*, vol. 30, no. 5, pp. 1073–1082, 2023. https://doi.org/10.1080/15376494.2022.2124002

[3] X. Cai, A. A. Malcolm, B. S. Wong, and Z. Fan, "Measurement and characterization of porosity in aluminium selective laser melting parts using X-ray CT," *Virtual Phys Prototyp*, vol. 10, no. 4, pp. 195–206, 2015. https://doi.org/10.1080/17452759.2015.1112412

[4] R. Gümrük and R. A. W. Mines, "Compressive behaviour of stainless steel micro-lattice structures," *Int J Mech Sci*, vol. 68, pp. 125–139, 2013. https://doi.org/10.1016/j.ijmecsci.2013.01.006

[5] E. Ptochos and G. Labeas, "Shear modulus determination of cuboid metallic open-lattice cellular structures by analytical, numerical and homogenisation methods," *Strain*, vol. 48, no. 5, pp. 415–429, 2012. https://doi.org/10.1111/j.1475-1305.2012.00837.x

[6] D. Tumino, A. Alaimo, G. Mantegna, C. Orlando, and S. Valvano, "Mechanical properties of BCC lattice cells with waved struts," *International Journal on Interactive Design and Manufacturing (IJIDeM)*, 2023. https://doi.org/10.1007/s12008-023-01359-9

Aerospace Science and Engineering - IV Aerospace PhD-Days
Materials Research Proceedings 42 (2024) 193-196

Materials Research Forum LLC
https://doi.org/10.21741/9781644903193-42

Extended reality-based human machine interface for drone monitoring in airport control tower environment

Marzia Corsi[1,a]

[1]Department of Industrial Engineering, Università di Bologna, Forlì, Italy

[a]marzia.corsi2@unibo.it

Keywords: Digital Technologies, Airport Control Tower, Air Traffic Control Operators, Advanced Air Mobility, XR-based HMI, Drone monitoring, Safety Net

Abstract. This study aims to exploit digital technologies to develop and validate an innovative *Extended Reality Human Machine Interface for drone monitoring in Airport Control Towers*. The model for the interface prototype exploits multimodal interaction features and implements technologies and functions such as Extended Reality, drone identification and tracking labels, and safety net visualisation.

Introduction

As unmanned aerial systems (UASs) continue undertaking a growing array of activities, the risk of encountering these vehicles is also rising. This poses potential threats to operational safety, including the possibility of collisions with other aircraft or structures and unauthorised entry into restricted areas like airports. Furthermore, the expanding scope of UAS operations necessitates increased investment and research into monitoring and managing drone activity technologies, including future "air taxis" [1].

A key challenge lies in identifying potential hazards posed by nearby drones. This study focuses on the concern of drones intruding into restricted airport zones, requiring airport safety and security units and control tower operators to monitor and manage such airspace intrusions. Various studies have explored methods for detecting unauthorised UAS activity and assessing the potential benefits of cooperative drone operations in the airport environment [2]. Recognising the need to monitor and regulate UAS traffic, Europe has developed the concept of *U-Space* under its Air Traffic Management research program *SESAR*. U-Space constitutes dedicated airspace for integrating UAS traffic with conventional aircraft, providing essential services for safe operations, and facilitating the identification and tracking of all involved actors.

Recent research has led to the publication of the fourth edition of the *U-Space Concept of Operations (ConOps)*, aligning with European regulations [3]. This document outlines requirements for safe drone operations, including identification, tracking, monitoring, geofencing, and segregation.

When dealing with Air Traffic Control (ATC) operations, monitoring UAS traffic near airports is expected to add additional workload to the controllers. As a matter of fact, supplementary dedicated interfaces intended for UAS traffic information shall be included in airport control towers. To improve ATC operations, while reducing controllers workload, several studies suggest to exploit digital technologies, such as eXtended Reality (XR), to provide users with innovative Human-Machine Interfaces (HMI) presenting surveillance information in a head-up display aligned with the controllers' direct view [4-5].

Leveraging insights from civil aviation traffic management, this work proposes extending XR visualisation techniques to Advanced Air Mobility (AAM) scenarios, drawing upon previous SESAR projects, such as *RETINA - Resilient Synthetic Vision for Advanced Control Tower Air Navigation Service Provision* and *DTT – Digital technologies for tower*. This study aims to

Aerospace Science and Engineering - III Aerospace PhD-Days   Materials Research Forum LLC
Materials Research Proceedings 42 (2024) 193-196   https://doi.org/10.21741/9781644903193-42

develop a novel XR-based interface that can be seamlessly integrated into control towers and to present unmanned traffic information in the airport vicinity for surveillance tasks.

XR-based HMI for drone monitoring in Airport Control Tower

Drawing from advancements in technologies such as those developed in projects like *RETINA* and *DTT*, this study endeavours to harness digital innovations to create and validate an XR-based Human Machine Interface tailored for Air Traffic Controllers (ATCOs) tasked with monitoring UAS traffic near airports. This interface aims to facilitate a more intuitive and efficient interaction within the control tower, enhancing both performance and situational awareness for ATCOs who must navigate the integration of autonomous drones.

Guidance from the U-Space concept of operations offers valuable insights into anticipated services, future information availability, and potential surveillance capabilities.

The interface design also draws inspiration from previous research on airport control towers, incorporating established concepts such as tracking labels, air gestures interaction [6], visual and aural cues for external elements, and safety net-based alert systems [7]. These inputs inform the development of a comprehensive interface tailored to the unique demands of monitoring UAS traffic in complex airspace environments.

In particular, the presented concept proposes including an on-demand semi-transparent augmented interface that live-streams the area surrounding the UAS ground infrastructure (vertiport) in the aerodrome area. To avoid the constant presence of an additional interface in the control tower, the ATCOs equipped with a see-through head-mounted display (ST-HMD) can visualise the HMI only when needed, exploiting the air gestures to activate and deactivate the interface (Figure 1).

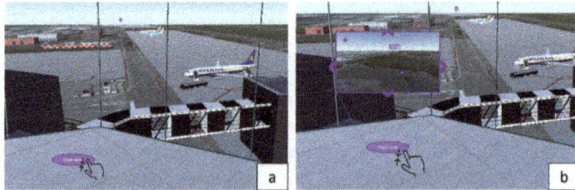

Figure 1 Proposed concept design for the XR-based HMI for drone monitoring in airport control towers.

After reviewing various sources, a suggested layout for visualised information emerges. This layout encompasses essential data points for the drones, such as identification details, velocity vector, battery status, ongoing activity, potential conflicts, and proximity to vertiport or airport boundaries. Several critical components are necessary for an effective surveillance interface, building upon interface designs crafted for airport control towers. These include a distinct marker to denote detected drone position, a corresponding label including the necessary surveillance data, and visual or auditory indicators to emphasise potential risks.

Assessment of the preliminary interface design

A human-in-the-loop validation activity is planned to be performed in a virtual environment to assess the goodness of the proposed concept and design. The user, provided with an HMD device, is immersed in a virtual representation of the Bologna Airport (LIPE) control tower. The simulation proposes to the user a futuristic AAM scenario where a UAS ground infrastructure is placed in an area of the aerodrome neighbouring the taxiway but outside the field of view of the ATCOs. The simulation foresees UAS traffic (air taxis) departing and landing from and to the vertiport.

Aerospace Science and Engineering - III Aerospace PhD-Days   Materials Research Forum LLC
Materials Research Proceedings 42 (2024) 193-196   https://doi.org/10.21741/9781644903193-42

Figure 2 XR-based HMI for drone monitoring proposed concept in a virtual environment representing an airport control tower.

Figure 2 shows a preliminary proposal for the interface design in a completely virtual environment. The three images show the active HMI displaying the vertiport close to the runway with a landing (green label), a departing air taxi drone (purple label), and the inactive interface. Figure 3 shows the representation of a safety net activated by a drone trespassing into a restricted airport zone. If a hazardous situation happens while the interface is closed, it automatically opens and signals the alarm through visual and auditory cues to guide the controller's attention. Moreover, the label of the involved drone turns red until the situation is solved.

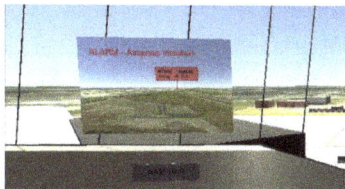

Figure 3 Safety net visualisation: a landing drone trespassing into a restricted area.

During the validation campaign, objective quantitative and subjective qualitative measurements from the ATCOs will be collected to refine the design and accommodate the user's needs before implementing the final interface to be assessed and validated in a real control tower. Figure 4 shows the proof of concept of the proposed HMI in a real airport control tower integrated with a real-time Augmented Reality (AR) interface for tracking airport traffic to support ATC operations,

Figure 4 XR-based HMI for drone monitoring (virtual) integrated with a real-time AR interface for tracking airport traffic in Bologna airport control tower as visualized by a user wearing a Microsoft HoloLens2 devices.

which is currently under development [8]. The interface streams a virtual scenario of a vertiport close to the runway and the related traffic and is provided to users wearing ST-HMD devices. In this real environment, the user can use the gesture interaction to display and deactivate the interface and move and scale it in the desired position.

Aerospace Science and Engineering - III Aerospace PhD-Days   Materials Research Forum LLC
Materials Research Proceedings 42 (2024) 193-196   https://doi.org/10.21741/9781644903193-42

Conclusions

The increasing prevalence of unmanned aerial systems in airspace operations necessitates effective strategies for monitoring and managing their traffic, particularly near airports where safety is paramount. This study explores the development of an XR-based Human Machine Interface tailored for Air Traffic Controllers to enhance situational awareness and operational efficiency. Drawing on insights from the U-Space ConOps and previous research on airport control towers, the proposed interface integrates essential features such as geoawareness, traffic information, and UAS identification. By leveraging XR technologies and incorporating elements like tracking labels and safety nets, the interface aims to provide ATCOs with intuitive tools for monitoring UAS activity. The interface design offers flexibility and adaptability to different operational contexts and allows for seamless interaction and visualisation of UAS traffic data. Moreover, features like on-demand interfaces and gesture-based interaction enhance usability and minimise cognitive load for ATCOs. The interface enhances safety and operational efficiency in managing UAS traffic near airports by providing timely alerts and visual cues for potential hazards.

It is possible to conclude that the development and validation of this XR-based interface can contribute to taking a step forward in addressing the challenges posed by integrating unmanned aerial systems into complex airspace environments. Through continued refinement and implementation, such technologies hold the potential to revolutionise ATC operations and ensure safe and efficient airspace management in the era of autonomous drones.

References

[1] Taha B, Shoufan A. Machine learning-based drone detection and classification: state-of-the-art in research, in IEEE Access, vol. 7, pp. 138669-138682, 2019. https://doi.org/10.1109/ACCESS.2019.2942944

[2] Heidger R, Lambercy V, Lambers D. Tracking analysis of drone detection systems at airports: methodology and results, 2021 21st International Radar Symposium (IRS), Berlin, Germany, pp. 1-17, 2021. https://doi.org/10.23919/IRS51887.2021.9466192

[3] U-Space ConOps and architecture (Edition 4), Deliverable 4.2 of SESAR 3 JU's Corus-Xuam project. 2023.

[4] Bagassi S, De Crescenzio F, Piastra S, Persiani C A, Ellejmi M, Groskreutz A R, Higuera J. Human-in the-loop evaluation of an augmented reality based interface for the airport control tower, in Computers in Industry, Volume 123, 2020. https://doi.org/10.1016/j.compind.2020.103291

[5] Santarelli R, Bagassi S, Corsi M, Teutsch J, Lasheras R, Carmona M, Groskreutz A. Towards a digital control tower: the use of augmented reality tools to innovate interaction modes, SESAR innovation days 2022, Budapest, 2022.

[6] Bagassi S, Corsi M, Galuppi F. V/A-R air gestures HMI interaction in airport control towers, ICAS 2022, Stockholm, 2022.

[7] Bagassi S., Corsi M., Extended Reality Safety Nets for Attention Guidance in Airport Control Towers, Aerospace Europe Conference 2023 Joint 10th EUCASS - 9th CEAS Conference, Lausanne, 9-13 July 2023.

[8] Fadda T, Bagassi S, Corsi M. ADS-B driven implementation of an augmented reality airport control tower platform, Materials Research Proceedings, Vol. 37, pp 767-770, 2023. https://doi.org/10.21741/9781644902813-164

Keyword Index